Senaa Abdullah Ali Al- jarjary
Raed Salem Ahmad ALsaffar
Mohamed Abdel-Raheem

Insects and forensic detection

Senaa Abdullah Ali Al- jarjary
Raed Salem Ahmad ALsaffar
Mohamed Abdel-Raheem

Insects and forensic detection

Noor Publishing

Imprint
Any brand names and product names mentioned in this book are subject to trademark, brand or patent protection and are trademarks or registered trademarks of their respective holders. The use of brand names, product names, common names, trade names, product descriptions etc. even without a particular marking in this work is in no way to be construed to mean that such names may be regarded as unrestricted in respect of trademark and brand protection legislation and could thus be used by anyone.

Cover image: www.ingimage.com

Publisher:
Noor Publishing
is a trademark of
Dodo Books Indian Ocean Ltd. and OmniScriptum S.R.L publishing group

120 High Road, East Finchley, London, N2 9ED, United Kingdom
Str. Armeneasca 28/1, office 1, Chisinau MD-2012, Republic of Moldova, Europe
Printed at: see last page
ISBN: 978-620-7-47841-5

Insects and forensic detection

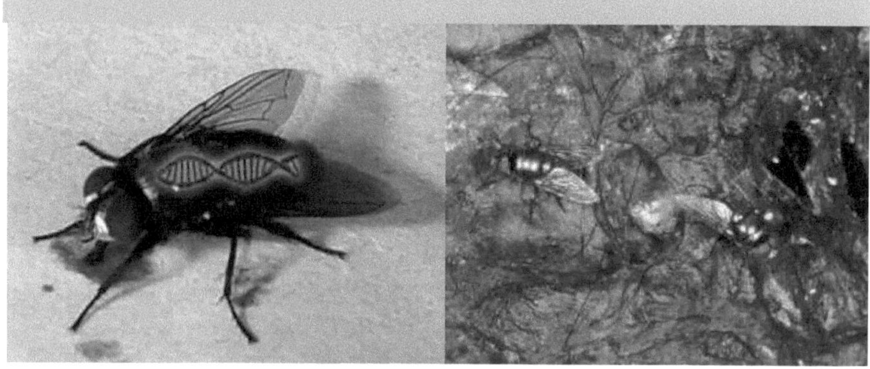

By

Assist. Prof. Dr. Senaa Abdullah Ali Al- jarjary

Prof. Dr. Raed Salem Ahmad ALsaffar

Prof. Dr. Mohamed Abdel-Raheem

(2024)

Insects and forensic detection

Assist. Prof. Dr. Senaa Abdullah Ali Al- jarjary

Department of biology, College of Science, University of Mosul. Iraq.

Email: Sensbio23@uomosul.edu.iq

Mobile: (+964)7701871500

Prof. Dr. Raed Salem Ahmad ALsaffar

Department of Forensic Evidence, College of science, University of Mosul, Iraq.

Email: raesbio2@uomosul.edu.iq

Mobile: (+964) 7701634647

Prof. Dr. Mohamed Abdel-Raheem Ali Abdel-Raheem

Professor of Entomology (Biological Control)

Pests & Plant Protection Department

Agricultural and Biological Research Institute

National Research Centre.

33rd ElBohouth St., Dokki, Cairo, Egypt.

Email: abdelraheem_nrc@hotmail.com,

abdelraheem_nrc@yahoo.com

Mobile: (+2) 01155527583 - (+*2) 01009580797*

Content list

INTRODUCTION

For centuries, insects have been assisting forensic scientists in solving crimes. The initial recorded event involved identifying a murderer of a farmer through the observation of blow flies attracted to the murder weapon (a sickle), ultimately leading to a confession. This particular event took place in 13th-century China. [1].

Forensic science continues to rely on important clues from the world of invertebrates. Below, you can find information on the types of insects used in forensic investigations and the reasons behind their use.

Forensic scientists rely on a variety of insects for their detection work, with an emphasis on flies (Diptera) and beetles (Coleoptera). Flies lay larvae that consume cadaver organs and tissues. As the corpse dries out, beetles take over the decomposition process. Other invertebrates, including those in the soil, also colonize the area around the cadaver.

Flies are crucial in forensic investigations, with Diptera species such as Calliphoridae (blow flies) being commonly involved. Blow flies are known for feeding on corpses and can arrive at the scene within minutes to hours of death. Examples of species include Chrysomya megacephala, Chrysomya rufifacies, Gaiijbhoria vicrna, and Lucilia sericata. Understanding the behavior and ecology of blow flies can aid in forensic detection. For instance, Lucilia illustris is typically found in bright, open habitats, while Phormia regina prefers shaded areas. [2].

The Sacrophagidae, also known as flesh flies, are similar to blow flies in that they feed on dead flesh and can quickly be found at a crime scene. House flies, of the Muscidae family, including species like Synthesiomyia nudesita, typically do not colonize a body until it has reached the bloat stage of decomposition. [3]

Skipper flies, fruit flies, and coffin flies, also infest the remaining tissues [2]. Various beetles such as carrion beetles, rove beetles, clown beetles, sap beetles, checkered beetles, scarab beetles, and dermestid beetles are beneficial in forensic examinations. These types of insects appear after the carrion or cadaver has started to dry out.

Ants, wasps, and species like springtails may also appear later on, either feeding on or utilizing a corpse or carrion as part of their environment [3].

The field of acarology focuses on the study of mites, ticks, and related organisms such as spiders. Forensic acarology specifically involves using information from these arachnids in criminal investigations, particularly those involving violent deaths. [4].

Various types of mites consume bacteria and fungi found in decaying matter, which then become the food source for other predatory mites. [5].

Mites can be beneficial to investigators in a number of ways.

They are capable of living on bodies and can be carried by visiting blow flies. Additionally, even storage mites like Tyrophagus longior have played a role in solving murder cases, while spiders have assisted in uncovering drug trafficking activities.

The diverse species of mites, due to their prevalence, have the potential to provide valuable information when necessary insects like blow flies are not present. One specific type of mite, Sarcoptes scabiei L., is noteworthy in cases of suspected neglect or abuse because it tends to infest individuals living in unhygienic conditions and unable to care for their personal hygiene. [4].

Insects found on a deceased body can give important information about the time since death, known as the Postmortem Interval (PMI), which can be helpful for determining the time interval between death and discovery of the corpse.

Estimating the age of immature insect stages feeding on a body and examining the species of insects present can help in determining postmortem intervals ranging from the initial day after death to several weeks.[6]

Researchers are looking into how insects and other small creatures found at a crime scene can help gather information such as location and climate, as well as details about the crime itself.

One area of interest is exploring the potential role of honey bees in forensic investigations:

1. There is a possibility that honey bees can be helpful in investigations related to wildlife trafficking[7]..

2. Scientists are investigating if honey bees can offer valuable information about missing individuals through the analysis of their honey[8]..

It has been stated that humans and insects are considered as the most successful creatures on Earth. Insects have been present on Earth for a much longer period than humans, approximately 350 million years prior to human existence which started around 100,000 years ago.

Humans strive to control their surroundings, enabling them to influence their success as a species. On the contrary, insects lack higher thinking abilities and tool usage, to the best of our knowledge.

However, insects possess the capability to adjust to different environments and lifestyles due to their small size, fast reproductive rate, tough exoskeleton, and the ability to fly when needed. These characteristics have contributed to making insects the dominant group of animals on Earth, both on land and in fresh water.

Insects' unique attributes make them valuable in the field of forensics. With the increasing popularity of forensics, many movies and popular TV shows like "CSI" are introducing young people to the techniques, tools, and critical thinking required to solve various crimes.

The emerging discipline of forensic entomology focuses on the information provided by insects to investigators concerning the timing, location, and potential criminal involvement in various criminal cases ranging from food contamination to homicide.

As it is a trending topic, many young individuals have witnessed the application of forensic entomology in movies and TV shows like "CSI" to assist in solving crimes such as murder cases.

A murder case always grabs the attention of young people, even if they are not up to date with the latest trends. It is interesting to note that the first time insects were used in a criminal investigation was in China in 1235. In this case, the murder of Chinese farmers was solved when flies gathered on a sickle used by the killer, revealing his identity through traces of blood and flesh despite his efforts to clean it.

The account was initially documented in China in 1247 in the book The Washing Way of Life, authored by Sung Tz'u, a lawyer and interrogator who detailed various human deaths he encountered and noted his findings.

Sung Tz'u also elaborated on the procedure for determining the likely cause of death, with the specific details he provided being considered fundamental principles that modern criminal anthropologists rely on, especially the absence of eggs from flies and maggots larvae.

In 1668, the renowned Italian physician Francisco L. Redi conducted significant scientific experiments that advanced the field of entomology and challenged the concept of spontaneous generation. Redi's studies aimed to refute the notion that fly larvae spontaneously develop from decaying meat under specific circumstances.

By examining rotting meat exposed to air and meat shielded from flies, Redi concluded that both fully exposed and partially exposed meat produced fly larvae. In contrast, the meat that was not exposed to the air did not attract any larvae.

The new finding not only disproved the theory of spontaneous generation, but also revolutionized the understanding of how living organisms decompose, leading to increased research in insect life cycles and the overall science of entomology.

In 1855, Dr. Bergeret d'Arbois used "criminal insects" in a criminal investigation, which was documented in a Case Report.

This marked the first instance of actual forensic entomology being applied to a criminal case. By collecting insects from a baby's body discovered in a house, it was determined that the insects indicated a state of decay dating back several years.

In his report, Dr. Bergeret d'Arbois detailed the life cycle of insects, made numerous hypotheses, and his methodologies closely resemble those still used in modern forensic entomology to estimate the time of death and the progression of insect species infesting bodies.

The first comprehensive investigation of forensic entomology was carried out in 1881 by the German physician Reinhard, significantly influencing the development of forensic science.

By studying multiple corpses, Reinhard was able to link the progression of insect species with bodies that had been buried.

His research laid the foundation for further studies and widespread research in the field.

In France, Megnin released a collection of articles discussing criminal entomology from 1883 to 1898. Following that, research in the area of entomology and its use in criminal investigations became more widespread, leading to the integration of criminal forensic entomologists in many countries worldwide today.

Lord and Stevenson, in 1986, categorized the study of criminal insects and its various uses into three main branches, which eventually became the fundamental fields of this science:

1- Urban Entomology focuses on insect infestations in human-constructed buildings and the surrounding environment.

2- Stored Products Entomology focuses on insect infestations in stored commodities such as grains.

3- Medico-criminal Entomology, also known as Medico-legal Entomology or Forensic Medico-legal Entomology, deals with analyzing insects present in cases of violent crimes like murder, suicide, and rape. In addition, a new field called Entomotoxicology has been developed to analyze toxins and drugs present in insect samples collected from corpses. Forensic Entomology, a scientific field, focuses on the impact of insecticides on body decomposition.

In criminal investigations, insect evidence is crucial for determining the precise time and location of a corpse. This information is unique to insects and cannot be obtained through other means. Furthermore, the study of insect behavior in Forensic Entomology utilizes principles from various branches of science:

- Taxonomy
- Ecology
- Toxicology
- Physiology
- Molecular Biology

To solve a crime, investigators must answer the five W'sǁ: *Who* was the victim, and/or the perpetrator; *what* happened; *when* did it happen; *where* did it happen; and *why* did it happen?

The Process of Death

Let's examine the concept of death in order to connect it to different approaches for determining the time it took place. It is important to understand that death is a gradual process, rather than a sudden event.

Various tissues and organisms within a living organism may die at different speeds. For instance, brain cells perish in 3-7 minutes when they lack oxygen, while skin cells can survive for up to 24 hours. The definition of death can differ between the legal and medical professions, particularly given the current scenario where individuals can be sustained by life support systems.

Physical Methods of Determining Time of Death

The time of death is typically determined by a pathologist, with the most accurate results being obtained if the body is discovered within the first 24 hours post-mortem, by examining algor mortis, livor mortis, and rigor mortis. If more than 24 hours have passed since death, alternative methods need to be utilized, with estimations based on environmental conditions and other scene-related information.

Algor mortis, which refers to the body's cooling process post-death, begins immediately after death as the body's temperature starts to equalize with its surroundings, following Newton's law of cooling which states that the rate of temperature change is directly proportional to the difference between the body's temperature and the ambient temperature.

Internal body temperature measurement can provide some insight into the time of death, as the body can no longer maintain its internal temperature of 98.6°F (37°C) after death and will begin to match the surrounding temperature. The process can be understood by considering Newton's law of cooling, which explains the relationship between temperature change and the temperature difference between the body and its environment.

This can be expressed mathematically by a simple first-order differential equation:

$$\frac{Dy}{DT} = ry \text{ or } dT\, dt = -K\,(T - TS)$$

This particular measuring technique was utilized for a period of time, however, it is only effective for small, non-living objects, making it suitable for hot coffee or a freshly baked pizza but not for deceased bodies. There are cases where the body temperature of a corpse remains constant for a period of time or even increases as a result of cellular reactions as they cease functioning, as well as due to heat generated by bacteria.

Simpler to apply than Newton's exponential temperature decay, and more accurate, is the Glaister equation:

$$\text{Hours since death} = \frac{98.4°\text{F} — \text{internal body temperature}}{1.5}$$

This formula can be applied within the time frame of one to 36 hours after death, however it is most reliable within the initial 12 hours. Normally, a body decreases in temperature by about 1-1 1/2 degrees Fahrenheit every hour until it equals the surrounding temperature.

Factors such as the environmental temperature, the clothing worn by the deceased, the level of moisture in the clothing, air circulation, number of clothing layers, and other variables must be taken into account. Additionally, individuals with a higher surface area to mass ratio, such as children or smaller adults, will cool at a faster rate.

Livor mortis is the accumulation of blood in the body due to gravity once the heart ceases to beat. This results in a purplish-red discoloration on the skin, which can provide insights into the body's position at the time of death.

Areas of the body that are in contact with the ground or constrained by other objects do not show livor mortis as the blood vessels in those regions are compressed.

The process of livor mortis starts within 30 minutes of death and is most pronounced within the initial 12 hours.

Following this period, the discoloration from livor mortis remains fixed and will not shift even if the body is disturbed, aiding in determining postmortem movement.

Rigor mortis is the stiffening of skeletal muscles that occurs after death. When a person dies, their muscles initially relax, but then ATP (adenosine triphosphate, an enzyme in the muscles) breaks down, fluid levels change, and the muscles become rigid.

Rigor mortis usually starts in the smaller muscles, so it is first noticeable in the face, neck, and jaw. This stiffness can develop within a few hours of death and typically disappears within around 30 hours, leaving the body loose.

The effects of rigor mortis fade away in the same sequence in which they appeared, with smaller muscles becoming loose first, followed by larger muscles in the trunk, arms, and legs. Environmental factors like temperature, dehydration, muscle condition, and activity prior to death can impact rigor mortis, making it difficult to accurately estimate the time of death based on this process.

Within four minutes of death, decomposition begins. Oxygen deprivation leads to an increase in carbon dioxide levels in the blood, a decrease in pH, and the accumulation of waste products, which harm cells. Enzymes then break down cells, causing them to rupture and release fluids, a process known as autolysis.

Autolysis is typically first noticed a few days after death when fluid-filled blisters appear on the skin and areas of the body experience skin slippage.

The next stage of decomposition, putrefaction, involves the breakdown of soft tissues, primarily by bacteria. A greenish hue to the skin is often the first visible sign of putrefaction.

The release of gases like methane and ammonia during ongoing tissue decomposition leads to bloating. The unpleasant smell is due to volatile acids like butyric and propionic acids.

Further breakdown of proteins and fats results in the formation of smelly compounds such as skatole, methyl disulfide, cadaverine, and putrescine.

At this point, the body is full of fluids and emits a strong odor. Some refer to this stage as "black putrefaction." The stages of decomposition are continuous, and the timing can vary based on different factors.

Adipocere formation may occur several months to years after death, resulting in a yellowish-white, waxy substance that forms due to the saponification of fatty acids, often catalyzed by a specific anaerobic bacterium.

The ideal conditions for adipocere formation include burial in a moist, alkaline soil (which slows down decomposition and prevents scavenging), high body fat content of the deceased, and burial in a casket inside a vault.

This process helps preserve the body, aiding in identification and injury recognition.

Mummification occurs when dehydrated tissues, typically skin, survive decay. This process commonly occurs in hot, dry environments with individuals who have low body fat, potentially starting as soon as a month after death.

In the last phase, the chemical composition of bones undergoes changes, primarily due to moisture and soil pH, a phenomenon known as diagenesis.

Various factors influence the decomposition rate, with the environment, temperature, and scavenger activity playing crucial roles. For instance, a corpse decomposes at twice the rate in water compared to underground, where it decomposes at half the rate.

The following equation gives a very rough estimate for time of decomposition of soft tissue for a person lying on the ground:

$$\text{Number of days to become skeletonized} = \frac{1{,}285}{\text{Average temperature, }^{\circ}\text{C}}$$

Therefore, if the average temperature near the body is 20°C (68°F), it will require approximately 64 days for only a skeleton to be left. In tropical areas, where the average temperature is 30°C, this process could happen in 30 days or even less due to the elevated humidity levels.

This information provides investigators with an approximate timeframe to start their investigation.

Criminal Applications of Entomology:

Establishing the moment of death is crucial when coming across a body (Post mortem Interval).

The duration elapsed since death can be ascertained, particularly with decomposing bodies. It is feasible to pinpoint whether the death occurred days, months, or with more precision, at night or during the day. This information is crucial in the prosecution or acquittal of any suspects.

Determine the actual location of death

Determining the physical location of the body can help determine the geographical distribution of insects in different environments when the body is found in a different location than the actual place of death, ultimately aiding in pinpointing the location of the crime.

Identify the culprit in Homicide murders

If an insect or part of it becomes trapped in clothing or tools, it can be seen as evidence of guilt. For instance, a priest was found guilty of murdering his wife after an ant from his shoe was determined to be of the same species as those found on his wife's body three days earlier.

Suicides

Sometimes the insect assembly of the developed or decomposed bodies may determine the location of the wound that caused the death / suicide, from which the type of suicide can be identified.

Causes of sudden death

The potential reasons for unexpected death, whether from toxicity or excessive intake of drugs or alcohol, are investigated by analyzing the insect's diet. This is where the same poisonous substance is discovered in the larvae consumed by the insect.

Blood Spatter Pattern

The movement of insects at the crime scene can alter bloodstains on surfaces as they come into contact with limbs contaminated with blood, vomit, or feces, potentially leading to incorrect conclusions about the blood distribution.

Carelessness of children or older

Cases of infection and lesions that have escalated to the point of attracting flies are examined to identify acts of negligence. This analysis can help determine the timing and severity of harm inflicted upon victims, including those who may still be alive.

Rape crimes

When examining the human DNA present in the insect's gut, it is possible to determine the identity of the individual as well as any possible criminal activity.

Identify the source of the smuggling of narcotic plants

Identifying the origin country of insect species can help in detecting drugs and tracking down their smuggling activities.

Location of Vehicle Travel:

Identifying the specific locations where the vehicle traveled by examining insects or parts stuck to the glass or radiator, using the identification of the insect and cross-referencing it with the variety and location of insects.

The importance of forensic science

Forensic medicine deals with solving mysteries related to crimes involving humans, such as torture, physical or psychological abuse, and murder.

Various methods are used to uncover these mysteries, including analyzing fingerprints, bite marks on the victim or suspect's body, examining dental characteristics like prosthetic teeth or deformities, and testing DNA found in teeth. Saliva left at the site of a bite or wound can also provide valuable information.

Lips also have unique characteristics that can be used for identification purposes.

Earprints, ear size, ear length, and the angle at which the ear makes contact with the head can vary among individuals and remain consistent throughout a person's life, unlike changes that may occur to the eyes or teeth.

Forensic osteology, a branch of forensic medicine, can help determine the age, sex, and genetic identity of a deceased individual, as well as identifying any body deformities. The length of a person's height can also be estimated by measuring bone length, particularly the femur.

In forensic medicine, Anthropology also focuses on the growth and development of human life. The origin and race of a person, whether Caucasian, African, Negroid, or Mongoloid, can be identified by examining the skull. For instance, African skulls have large eyes that are rectangular or square, and the nose bone is not pointed. In contrast, European skulls have round eye sockets and a pointed nose.

The gender of an individual can be distinguished by the shape of the skull, with male skulls being pointed at the base where it connects to the neck. Female skull features include a square chinbone, a flat forehead bone, a Sup-Pubic pubic bone, and wider pelvic bones compared to males.

Furthermore, signs of torture on a person who may have been detained can be identified by the presence of Myiasis symptoms, such as ulcers appearing in various locations on the body, particularly around natural openings if the individual's hands were restrained and unable to shoo away flies.

In cases of shooting incidents, forensic experts can determine the trajectory of the bullet within the body to pinpoint its location. Parodic freckles on the victim's body can indicate that the shooting occurred in close proximity to the person, possibly within the body itself.

In the field of forensic medicine, facial reconstruction can be performed using special computer programs to create a likeness of the actual face in cases of deformities caused by terrorist incidents. This helps in identifying the individual through facial recognition.

Forensic Medicine also covers the study of death, known as Thanatology, which involves the physical changes that occur after death such as rigidity of the body, lactic acid buildup, eye and leg erosion, cessation of heartbeat, muscle relaxation and stiffness depending on the time after death.

The moments leading up to death are characterized by sensory perception without response, as the person is still conscious but unable to communicate. This is due to increased alcohol levels in the brain caused by dehydration and lack of oxygen, affecting consciousness before death.

In the Holy Quran, death is described as a natural process in Surat Q - verse 19, where Allah mentions death scars.

"The truth of death came from the truth, so that I would not deviate from it."

حيث قال الله تعالى" وجاءت سكرة الموت بالحق ذلك ما كنت منه تحيد" صدق الله العظيم.

Forensic Entomology

- **The role of insects in human life**

Millions of years ago, insects were present before the arrival of humans, with the collimbola insect being discovered in Scotland at the same time.

Insect species evolved, while others became extinct, leading to a total of 29 ranks of insects comprising 80% of geological group types. Insects are the most widespread organisms on Earth, possessing the ability to move through jumping, flying, or crawling.

Insects are adaptable to various climatic conditions, including cold or tropical environments, with some being active during the day and others at night.

Wings allow insects to search for food and locate suitable places to lay eggs, with flies and beetles being among the most economically important insect species.

There are around 300,000 registered fly species out of an estimated 750,000 species worldwide.

Insects undergo several growth stages from egg to adulthood, with varying durations for each species depending on environmental conditions.

The time required for insect growth typically increases at lower temperatures, highlighting the significance of temperature control at crime scenes.

- **The Holy Quran first referred to the role of criminal insects.**

The tale of our master Solomon, may peace be upon him, involves the teaching of bird language and insect logic by God. According to God, when Solomon's death was decreed, the earth consumed his body but his senses remained intact. The other beings realized that they do not possess knowledge of the unseen and feared God's punishment. This incident is mentioned in the 14th verse of Al-Sabaa.

قال الله تعالى " فلما قضينا عليه الموت ما دلهم على موته إلا دابة الأرض تأكل منساته فلما خر تبينت الجن أن لو كانوا يعلمون الغيب ما لبثوا فى العذاب المهين" صدق الله العظيم سورة سبأ الآية 14

Forensic Science

Insects live and coexist with humans throughout their life, often found in various areas of the body, particularly those with hair or on the skin, leading to certain diseases.

After death, insects are drawn to the body, especially flies, which are attracted to the body's natural openings such as the eyes, mouth, ears, nose, throat, and genital area, where there is suitable moisture for laying eggs and growing fly larvae. This is illustrated in Fig. (1).

The significance of certain types of flies lies in their ability to help determine the time of death during the initial stages of body decomposition, particularly within the first and second weeks, by estimating the age of larvae present with the body.

Flies are distinct in their ability to locate and be drawn to a body after death for a brief period, even if the body is in an exposed location or buried underground. They are guided by the scent of decay emitted by the body as it undergoes decomposition

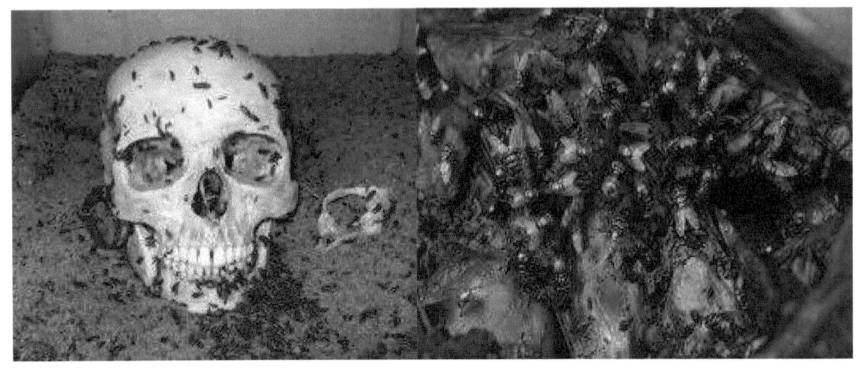

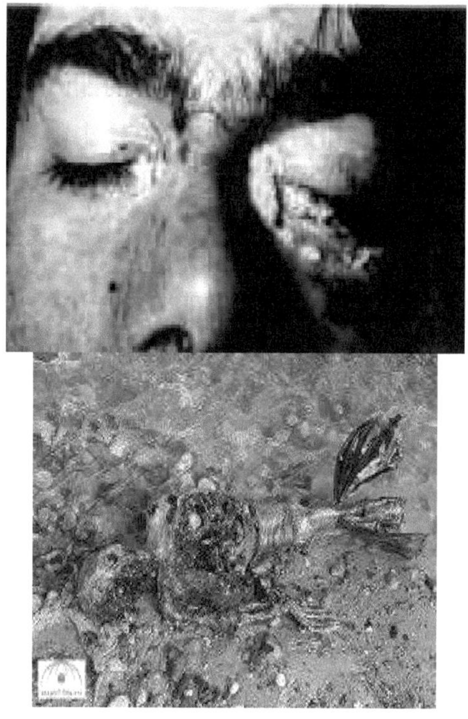

Fig. (1): After death, white blocks can be found in the eyes and flies may be present in various openings of the body .

In 1668, Italian physician Francesco Redi discredited the theory of insects self-creating, which suggested that worms on decaying meat emerged independently and proved that larvae develop into flies, thus impacting public health in the 1890s.

Scientist Megnin was the first to document the various types of insects found on human bodies in 1894, providing a way to determine the age of a body based on these insects.

The blue flies from the Calliphoridae family are crucial in forensic entomology, feeding on a corpse immediately after death, revealing crucial information in criminal investigations. Sung Tzu, a Chinese scientist, recounted a story in his book Mistakes about a murder case where villagers were asked to bring their machetes, leading to a confession when flies were found on a particular machete with blood traces. This revelation guided justice to the killer

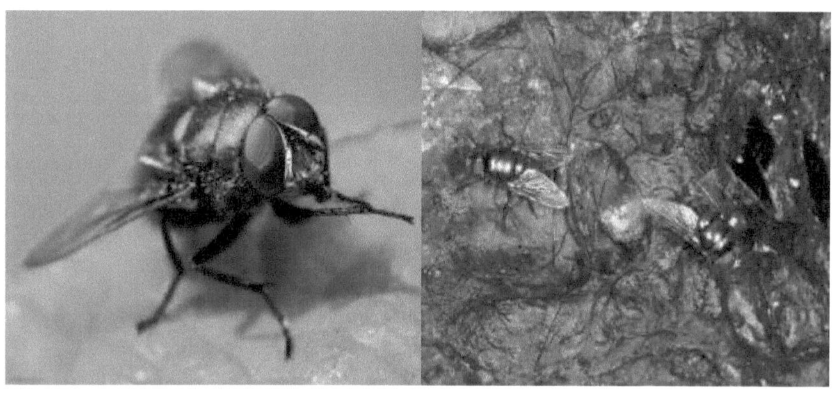

Fig. (2): Adult blue flies, as well as pupae and larvae, make up the three life stages of this insect.

In 1825, the French scientist Bergret discussed the idea of insects appearing on a body after death, and in 1894, the French scientist Mignin explored the connection between body decomposition and the arrival of waves of insects, which can be used to estimate the time of death.

In 1993, American scientist Le Goff categorized the insect species found on bodies into four groups based on their functions and food preferences:

1-Necrophagous

These bugs consume the corpse and are a crucial criminal group that plays a significant role in determining the Post-Mortem Interval (PMI), particularly within the first two weeks after death. The bugs consist of flies and beetles.

2-Predators and Parasites

These insects are known for feeding on other insects that are considered to be of high criminal importance. This group consists of beetles and certain types of fly larvae, which initially feed on the body tissue of their prey before becoming predators as they mature.

3- Omnivorous

They are bugs that consume dead animals and other bugs, such as wasps, ants, and certain types of beetles.

4. Incidental insects

They are insects that view the corpse as an additional food source in addition to being present in the soil like insects such as colimbula, spiders, dream, and mother of forty four. This group primarily consumes fungi that develop on the body.

The role of insects in detecting crime

In criminal investigations, insects are essential for detecting crime, identifying perpetrators, and even proving the innocence of the accused to ensure justice:

1. Determining the date of death when the body is found:

Flies are utilized in different forms throughout the various stages of decomposition, whether it be as larvae, pupae, or as part of a succession of insect species that are attracted to the body.

• Use of larvae in determining the date of death:

The age of larvae found on a carcass can be determined by measuring the size of the largest larvae and calculating the time since hatching, taking into account the temperature at the crime scene. This method is most reliable within the first three days of death.

For example

In 1994, a woman was murdered in Oklahoma, USA, after a violent argument with her husband. The husband had reported her missing a day before, and it was discovered that she had been buried in the backyard. When the body was discovered, forensic experts noted the presence of insects that had been there for three days.

Another example

In southern England, a corpse was discovered with extensive burn injuries. The forensic doctor faced challenges in establishing the time of death, however, the insect specialist managed to calculate the age of the larvae found near the body. The larvae were determined to be six days old, indicating that the fly eggs had been deposited on the body a day prior. The fire that caused the burns was observed in the house a week ago on the same night, as determined by the insect expert's analysis of the fly larvae's age.

• The use of the Pupa phase in determining the date of death:

The duration of the fly's life cycle is determined by calculating the time from hatching to reaching the full insect stage after emerging from the pupa. Detectives can collect the pupa or pre-pupa stage from the crime scene and use it to establish the date when the fly left the pupa.

• Calculate the date of death according to the succession of different insects on the body

In situations where the body has been left for three months or longer before discovery, various insects such as beetles, flies, ants, and mites are found in different stages. Research has demonstrated a strong connection between the five stages of body decomposition and the types of insects present in the crime scene during different weather conditions.

For example

In 1850, a child's body was discovered inside a chimney during renovations in Paris. The body was analyzed by the French medical specialist Marcel Bergret, who found larvae from the meat fly and certain butterflies on it. Marcel Bergret concluded that the body had been wrapped in 1848, with butterflies gaining access to it in 1849. Consequently, the individuals currently residing in the house were deemed innocent, since they arrived after the butterflies had already entered the body. The blame for the murder of the previous occupants was placed on the child.

2. Discovering the person responsible for the crime:

Insects play a role in identifying the perpetrator by examining the insects discovered at the crime scene and on the body to establish the human DNA.

Example

In 2008, a deceased driver was discovered in his vehicle in Finland. When authorities investigated, they discovered a mosquito with blood on it inside the car. This mosquito was preserved in a container and sent to a forensic lab to analyze the DNA of the blood.

The DNA matched that of a human, and it was compared to the genetic profile of a suspect. This evidence ultimately led to the suspect confessing to the crime, as the mosquito placed them at the scene of the crime when examined.

3. Identifying the location of the crime scene:

Insects have varied habitats based on their surroundings, coexisting with other organisms without interference. Some insects reside in water and air, while others have adapted to different habitats in rural and urban areas.

Example

In 1935, the bodies of two women were discovered in a car submerged in a river in Edinburgh, Scotland. These bodies were infested with fly larvae in the advanced stage of development.

Flies are typically found in soil, not water, indicating that the deaths occurred on land rather than through drowning in the river. It was deduced that the victims were killed elsewhere and then placed in the car, which was subsequently dumped in the river to make the deaths appear accidental.

Further investigation revealed that the victims were a woman and her servant, with the culprit being identified as the husband of the woman, Dr. Buckston Ruxton.

4. Identification of illegal drug transportation:

Insects play a role in identifying drug smuggling by categorizing the types of insects that are linked to the drugs. This allows for the identification of where the drugs originated from and provides proof of their smuggling. The types of insects found can differ depending on their geographic location.

Example

In 2005, a shipment of marijuana that was being smuggled into the United States was intercepted. When examined, only officially recorded spiders from

Colombia were found, suggesting that the drugs had been smuggled across the border with Colombia.

Similarly, in New Zealand, a drug trafficking case involved a shipment of cannabis containing three species: Tenebrionidae, Bruchidae and Carabidae. This shipment also included some species of wasps and ants.

Research into the geographical distribution of these insect species revealed that the beetles found with the cannabis were only known to inhabit the Temmaserin region near the Thai border. This information led to the confession of one of the suspects involved in the smuggling operation, based on the findings of entomologists analyzing the insects linked to the cannabis shipment.

5. The identification of harmful substances and medications within the bod:

Substances like heroin and cocaine impact the development of fly larvae, causing them to grow faster. When the larvae are fed tissues contaminated with a pesticide like malathion, their growth is slowed down. The chemical examination of the larvae extract from the body can determine if the individual consumed a harmful substance intentionally for suicide or as a drug addict or under the influence of alcohol, especially in cases where the body is badly decomposed making it hard to analyze chemically due to the breakdown of toxin residues. Various analytical methods are typically employed for this purpose:
Gas Chromatography or Tin layer Chromatography in such a case as an

Example 1:

In 2005, a man's body was found in Honolulu, France under unexpected circumstances, with a history of attempted suicides. The man's body was discovered next to a bottle of malathion pesticide, a toxic phosphorus compound.

Analysis of larvae found on the body revealed the presence of the same compound, indicating that death occurred five days prior. However, the man had been missing for about eight days before the body was found.

Further analysis showed that the lack of fully developed insects around the body was due to the negative impact of Malathion on fly larvae growth, resulting in premature death before completing their life cycle. This explains the discrepancy between the time of disappearance and death, as estimated by the dead fly larvae that did not reach full development.

Example 2:

In 2007, the body of a man was discovered in an abandoned house in Torren, Italy. The man was found wearing multiple layers of clothing and surrounded by insects, indicating that the body was in an advanced state of decomposition, making it challenging to perform a chemical analysis.

Initial findings suggested that the man's body may have gone missing from a nursing home around the time when the body was found. The absence of internal bleeding on the body suggested radiation exposure, while analysis of his hair confirmed his alcohol addiction. Additionally, the chemical analysis of larvae extract indicated that the cause of death was alcohol poisoning.

Example 3:

In 2004, the remains of a 29-year-old man were found in the woods. He had been missing for several days and was believed to have consumed cocaine.

Upon examination, there were no signs of external injuries on the body and no drugs were present. Larvae near the body were collected for further investigation.

The larvae and body samples were sent for analysis, which revealed the presence of cocaine in both. It was determined that the man had died from a cocaine overdose.

6 - Identifying acts of concealment by altering position and relocating:

Serbian soldiers killed around 7,000 Albanian Muslims in Kosovo and subsequently executed prisoners of war and wounded soldiers in hospitals before burying them in mass graves.

When the graves were found, it was observed that insects linked to the corpses were from various geographic locations, suggesting the killings happened in different regions.

The bodies were gathered and concealed in mass graves to conceal them and prevent ethnic cleansing or genocide.

Example 1:

In 1962, a deceased individual was discovered in a saltwater tank in Azerbaijan, with live fly larvae found near the body. The larvae were not typically found in saltwater environments, as flies are typically ground insects rather than aquatic.

This suggests that the body had recently been moved to the tank with the fly larvae. Analysis of the fly larvae indicated that the individual had been killed nine days prior to the body's discovery. The perpetrator confessed to initially concealing the body in another location before disposing of it in the tank.

Example 2:

In a recent book titled "A Fly in Prosecution" by Lee Goff, an American expert in criminal entomology, the author explores the significance of insects in solving murder cases.

The story revolves around the discovery of a woman's body on a sugarcane plantation in Ohio, USA. According to the findings of the criminal insect specialist, certain species of flies found on the body were typically associated with urban environments rather than agricultural areas.

Through police investigation, it was determined that the killer had stored the body in a city residence before transporting it to the farm. This allowed for the presence of urban flies to infest the body, laying eggs and multiplying, which served as crucial evidence of the crime scene.

Example 3:

In 2007, two girls' bodies were discovered in a forest and grassy area in southern Italy. They had been placed in plastic bags.

One of the bodies appeared to be in an advanced stage of decomposition with a significant number of egg flies present, while the other body had fly larvae and eggs. The flies found on both bodies were examined in a laboratory, revealing that the fly eggs on the first body belonged to two different types of flies, one of which had never been seen in that particular agricultural area before.

Similarly, the flies found on the second body were also unknown in the region. This suggested that the girls had been killed elsewhere, as the flies present on the bodies were in various stages of development. It was concluded that the bodies were likely moved to the forest to throw off investigators.

This tactic was discovered as both bodies were infested with a type of fly that was previously unrecorded in that area, as shown in Figure 3.

Fig. (3): Different stage of flies

7. Revelation of acts of torture and abuse:

Several insects, like flies, including Blow flies, play a role in revealing torture crimes happening in detention centers when detainees are restrained and unable to fend off the flies. This is especially true when detainees are injured and unable to clean themselves.

Injured areas are appealing to flies for laying eggs and larvae, leading to sores and patches known as Myiasis. Flies are attracted to body openings such as the mouth, ears, nose, eyes, and genital area. (Fig. 4).

Older individuals, particularly those experiencing incontinence, and children with disabilities who are not receiving adequate care are at a higher risk of developing skin infections caused by flies being attracted to the scent of ammonia from urine and laying larvae on the skin.

It is important to recognize that neglecting these symptoms in the elderly and children can lead to allegations of family violence against those responsible for their care.

Example 1:

In 1634 in Germany, a woman went out of her house after drinking wine and left her child alone at home. Sadly, the child was discovered dead in his room.

During the investigation, larvae were discovered at the bottom of the child's diaper, resulting in him being infected with maggots. These larvae were from the third instar stage of the Musca stabulns species, as well as from the small domestic fly Farina canicularis.

The blow fly Callifora vomitoria was found on the child's face. The woman's neglect and lack of care for her child's well-being left him in poor health, ultimately leading to his death. This negligence led to her facing charges for failing to properly care for her child.

Fig. (4): Blow fly life cycle and chasing flies for wounds

Example 2

In 2002, a woman's body was discovered in her home in Germany, where domestic flies and beetles were found on her body, which typically indicates exposure to urine and feces due to lack of proper health care for the elderly and children.

Despite the presence of flies, there were no signs of larvae feeding on the body, and her eyes were in good condition. Ulcers were observed on her foot, which was enclosed in a tightly bound plastic bag infested with fly larvae estimated to be four days old.

It was determined that the foot injury and fly infestation occurred two days prior to the woman's death, suggesting she had been neglected by her caregiver, Abha, who was accused of neglecting her mother's care.

8. Discovering instances of sexual assault and determining the identity of the perpetrator

Research has indicated that it can be challenging to collect semen samples from the body of a sexually assaulted female in advanced decomposition to identify the genetic material of the attacker. However, larvae found in the genital area could potentially provide important information on whether the female was raped before her death, as these larvae may carry the genetic signature of the rapist.

This suggests that the genetic material of the perpetrator can be transferred to fly larvae feeding on the victim's body through semen. As a result, it is crucial to isolate and preserve the fly larvae located in the genital area of the female body in alcohol, ensuring that the contents of their intestines, which might contain victim tissues and the perpetrator's sperm, are not lost.

9. Identifying terrorist activities and acts of war crimes

The larvae of flies discovered on the bodies of victims killed in explosions have been found to contain traces of the explosive materials used in the terrorist attacks as they feed on the remains. Analyzing the contents of these larvae can provide insights into the type of explosives used in the attacks.

Comparing the chemical composition of these larvae with that of explosives used in previous incidents can help determine if the same materials were used, potentially indicating whether the attacks were carried out by the same terrorist group or multiple groups.

Insects found on bodies, particularly in mass graves, can help identify the origin of victims of war crimes. By examining the types of insects present, authorities can determine if the victims were prisoners of war taken from various locations and executed then buried in mass graves.

Insects can serve as valuable tools for the International Criminal Court (ICC) in investigating and prosecuting war crimes.

Example

In 1983, a Korean civilian aircraft entered Soviet Union airspace and was subsequently shot down by a missile. Flies' larvae found on the victims' remains from the plane contained traces of TNT, indicating the use of the missile in the attack.

Thus, the fly larvae were evidence that the plane had been shot down by a missile and not as a result of a mechanical failure that hit the plane and therefore a war crime.

It is certain that intelligence agencies from various countries may seek assistance from forensic experts in the future to solve the mysteries related to numerous crimes and terrorism problems.

10. The impact of drugs and poisons on the development of fly larvae and how it influences the time of death.

Insect larvae can store body tissues from victims who may have overdosed on drugs, leading to death. Research has shown that larvae fed meat with narcotic substances experience delayed development, while those fed drug-free meat develop faster.

Drugs like heroin can extend the pupal stage, potentially delaying the estimation of time of death by around 24 hours if larvae contain drug remnants. Therefore, insects can be used to estimate the time of death in cases where the deceased may have been drug users. Sure, please provide me with the text that you would like to be paraphrased.

Gathering insect specimens with forensic significance from the location of a crime

1. The tools required for collecting samples:

To collect flies with the body or birds above, a manual net and glass tubes with 70% ethyl alcohol are needed for laboratory classification.

Additional glass containers are necessary to house live specimens of fly larvae and other insect species for breeding in the laboratory. It is important to also monitor the temperature of the atmosphere both under the body and above the surface, as well as the atmospheric humidity.

2. Steps to collect insects accompanying the body before being transferred from the scene of the crime:

- Photography is necessary in order to document the body and all relevant information should be recorded with the support of images.
- Flies should not be disturbed prior to collection in order to preserve them for classification purposes.
- Some fly larvae should be collected and kept alive to observe the complete life cycle.
- Larger fly larvae should be collected and preserved in alcohol-filled glass tubes for measurements to estimate age and time of death.
- Flies from the soil beneath the body should be collected and kept alive in containers for further classification in the laboratory.
- Insects crawling on the body, such as beetles, should be collected individually for classification and to determine the stage of body decomposition.
- If the body is female, fly larvae in the genital area should be collected to analyze DNA and determine if there was rape before death.

- Larvae on wounds should be collected separately for laboratory examination to determine injury timing.
- Dead insects found with or below the body should be collected and stored separately.

3. Collection of insect samples from the scene of the crime after the transfer of the body:

When a body is found in an open area and then moved from the scene, there will be various insects and arthropods at different stages present. It is important to adhere to proper procedures for collecting samples from the crime scene and preserving other specimens for further study on a farm.

Fallen tree leaves and other materials near the body should be collected along with relevant data, and these samples should be taken to the laboratory for analysis. Soil samples up to 3 feet deep are collected from the body's location, stored in paper bags, and documented for examination.

4 - Collection of insect samples accompanying the body after the transfer of the body to the morgue:

After relocating the deceased to the morgue, it is recommended to gather the insects discovered on the body in the deeper crevices. The body is typically transported in a fabric sack. Once the body is taken out of the bag, the interior of the bag should be inspected, as insects and arthropods often emerge from the body due to discomfort.

Upon placing the body on the examination table, it is essential to obtain additional samples of insects from the body, particularly from the clothing folds where they tend to lay eggs, house larvae, feed pupae, and occasionally, dwell in certain instances.

5 - Collection of insects associated with a body found in a tightly closed place:

The confined space surrounding the dead body presents a challenge for forensic professionals, particularly when access to the location, such as a vehicle or residence, is restricted .

Efforts are made to prevent the odor of the corpse from escaping and attracting insects .

The body is shielded from insects in the vicinity and as a result, it becomes difficult to gather necessary information or draw conclusions that could shed light on the circumstances of the crime.

6. Preparation of the report of the criminal insect expert:

The criminal insect expert must prepare a detailed report for submission to the court in order to identify the circumstances of the crime. The report must include all data, conclusions, and estimates related to the crime to avoid the need for the expert to explain the report.

Evidence supporting the report's findings must be provided, along with observations from the crime scene such as whether it occurred in an urban or rural area, the state of the body (clothed or naked, injured or not), and whether the body was buried or left on the ground. Factors like the body's decomposition and the weather conditions at the time of discovery should also be mentioned.

Specific details about the types of insects found on or near the body, as well as how they relate to the report's conclusions, should be included along with photographs of the crime scene and the body before and after removal.

The report should specify when insect samples were collected and when they were sent for laboratory analysis. It should also cite scientific references that were used to support the conclusions presented in the report.

Additionally, the report should include information about the expert's background and experience in the field, including any recent scientific conferences attended. This is important to establish the expert's credibility and expertise in the subject matter.

It is important to emphasize that the report's purpose is not to incriminate anyone, but rather to uncover the truth and dispel any uncertainties surrounding the crime in order to achieve justice.

Fauna Succession Chart (Joyce Chan, 2012)

Table (1): Fauna Succession Chart (Joyce Chan, 2012)

Waves of insects	The main insects	Condition of the body	Age of the body
The first	blowfly	Fresh	1-2 days
The second	flesh fly	Surge odor	The first three months
The third	Dermestid beetles	Fat out of the body	3-6 months
The fourth	Different flies		
The fifth	Flies and beetles	Surge odor	4-8 months
The sixth	Mites		6-12 months
The seventh	Dermestid beetles	Completely dry	1-3 years
The eighth	Different beetles		3 years and more

Major Insects Families of Forensic Interest

Table (2): Major Insects Families of Forensic Interest

Fly Families	Beetle Families
Califoridae	Silphidae
Sacrophagidae	Dermistidae
Piophilidae	Staphylinidae
Sphaeroceridae	Histeridae
Scathophagidae	Cleridae
Sepsidae	Trogidae
Fannidae	Carabidae
Phoridae	Scarabaeidae
Muxcidae	Nitidulidae
Muscidae	

Varieties of flies that are significant in criminal investigations, along with their life cycles and reproduction processes in a controlled environment

1- Types of criminal flies:

Flies are regarded as the most crucial insects in the realm of criminal investigations, being the first insects to gather around a dead body and lay eggs on it. These eggs eventually hatch into larvae that play a significant role in helping to solve crimes and unravel mysteries.

Belonging to the Diptera order, flies are winged insects with over 70,000 different species. They are distinguishable from other insects by having only one pair of wings, with the second pair transformed into small organs known as halters.

Diurnal flies are active during the day and rest at night, with many species feeding on floral nectar or organic matter. The larvae, on the other hand, are either predatory or parasitic, while certain adult flies consume the blood of vertebrate animals.

- **Calliphoridae flies**

This particular species closely follows three major types of flies that are significant for medical and criminal purposes, commonly referred to as blowflies or bottle-flies.

The flies within this group have a tendency to avoid laying their eggs in the darkness, particularly in rural regions without any artificial lighting, and this species is known to closely follow numerous other species.

1-Fly *Calliphora vicina*

It is a sizable fly with a body measuring approximately one centimeter, displaying a blue-silver hue (Figure 5)

Calliphora vicina

Fig. (5): ***Calliphora vicina***

2- *Calliphora vomitaora*

Fig.(6) depicts a big blue fly that has a longer life cycle compared to the one before it. It is typically found in rural locations.

Fig. (6): ***Calliphora vomitaora***

3- *Lucilia sericata*

The green metallic color of these flies is what distinguishes them, which is why they are known as Greenbottle flies, as shown in Fig. (7).

Fig. (7): ***Lucilia sericata***

4- *Lucilia illustris*

This particular insect holds significant importance in forensic investigations, particularly in Washington, DC, where it was involved in a case that highlighted the ambiguity of murder in 1986, as depicted in Figure 8.

Fig. (8): ***Lucilia illustris***

5- *Phormia regima*

Its size is smaller than the size of the previous species of flies, with a length of less than one centimeter.

The color of the body is olive green and the color of its head is black and is therefore known as the black blowfly, Fig. (9).

Fig. (9): *Phormia regima*

- **Phoridae flies**

The flies of this very small species have a fly length of 1.5-6 mm and brown or blue gray.

These flies are characterized by their ability to run and jump. The species includes more than 2,500 species, some of which are predators, while others are parasitic, Fig. (10).

Fig. (10): Phoridae flies

• Piophilidae flies

The cheese skipper flies, which feed on animal products and fungi, typically follow the mentioned species.

Among these flies, the most common one is the piophila casei. This small fly, measuring less than half a centimeter in length, typically feeds on meat, fish, cheese, and dccomposcd animals.

These flies usually appear on bodies approximately 3-6 months after death, during the decomposition stage transitioning into the dry phase.

The fly can be identified by its black or blue color, with yellow markings on the head and legs, and is about half the size of a common housefly.

The larvae of these flies have the unique ability to jump up to one centimeter in the air when disturbed. Additionally, the size of the larvae exceeds that of the

adult flies.

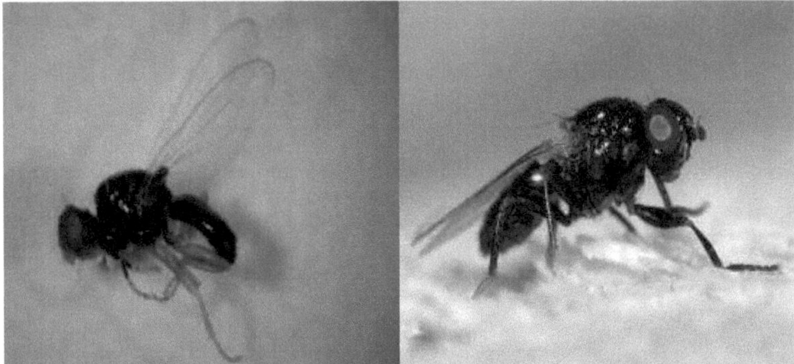

Fig. (11): Piophilidae flies

• Muscidae flies

This particular group of insects consists of common houseflies, specifically the Musca domestica species, which are drawn to human bodies after death due to the scent of bodily secretions, particularly from natural body openings. These flies are typically gray in color and measure 7 millimeters in length.

The female flies lay eggs in clusters of 100-150 on organic material, with the larvae developing in 5-14 days before moving to the soil surface to pupate, ultimately emerging as mature adult flies in 3-10 days depending on temperature. Commonly found year-round, these flies are particularly prevalent in the spring and summer months .

Fig. (12): Muscidae flies

• The stable – fly

This fly belongs to the same species as the previous Muscidae and is scientifically known as Stomoxys cucitrans. It is also known by several other names such as Barn fly, biting fly, or dog fly.

Unlike other species within the same group, this fly feeds on mammalian blood. Its appearance is similar to that of a house fly, but it is smaller in size, with a body marked by distinctive spots.

The fly is less than one centimeter in length and its color tends to be lighter than that of a house fly. It is often found in areas with manure, such as livestock pens and bird cages.

This fly is a carrier of numerous animal diseases. Both male and female flies feed on mammalian blood during the day. Males typically die after mating, while females die after laying eggs. The life cycle of this fly lasts more than 30 days.

Fig. (13): *Stomoxys cucitrans*

- **Sphaeroceridae flies**

The small dung fly is a type of fly that is gray in color and measures between 0.5 to 1.5 centimeters in length.

This species of fly is drawn to the smell of manure and decomposing bodies.

There have only been a few instances of this species being found on dead bodies, as shown in figure 14.

Fig. (14): Sphaeroceridae flies

• Fannidae flies

Flies of this kind are known as latrine flies, and they are smaller in size compared to other flies. They are drawn to light. In Germany, Fannia canicular flies were found in a situation where a child was neglected. These flies are attracted to the scent of urine and feces. Their eggs develop into larvae that can cause ulcers and infections in the genital area, Fig. (15).

Fig. (15): Fannidae flies

•Sacrophagidae flies

Flies known as Flesh flies, where the larvae of this species feed on decomposing tissues of human and animal bodies, as well as living tissues causing various diseases such as ulcers and infections. These meat flies are characterized by their large size and gray-silver color, with the most common species being Sarcopha bercae.

It is important to note that the female flies of this species, known as Lodeh, lay eggs inside themselves and can lay a number of 40-80 larvae, which then live on organic matter and decaying flesh. These larvae may be found in man's sinuses or on infected wounds, Fig. (16).

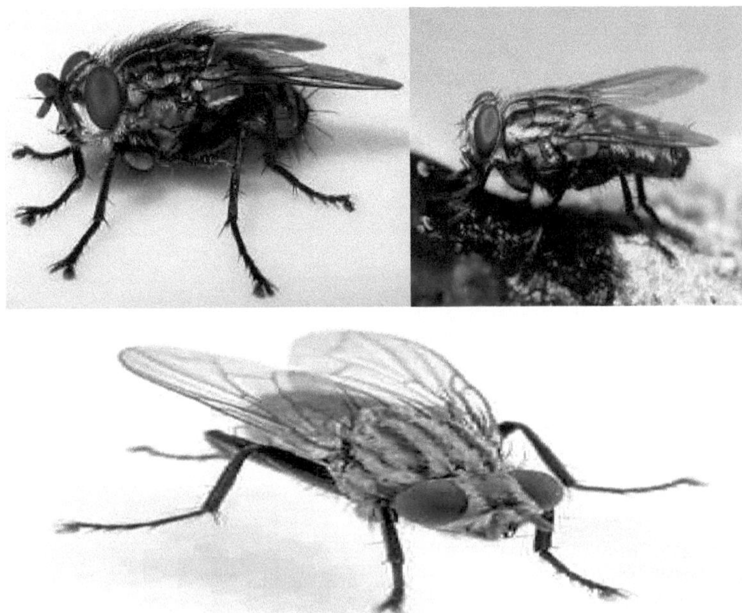

Fig. (16): Sacrophagidae flies

Life cycle of flies

The life cycle of flies starts with the female egg hatching into larvae, which then matures into a pupa and eventually a full-grown insect.

Certain fly species develop within the female's body, with females giving birth to larvae like in the case of the meat fly, known as viviparous females (Fig. 17).

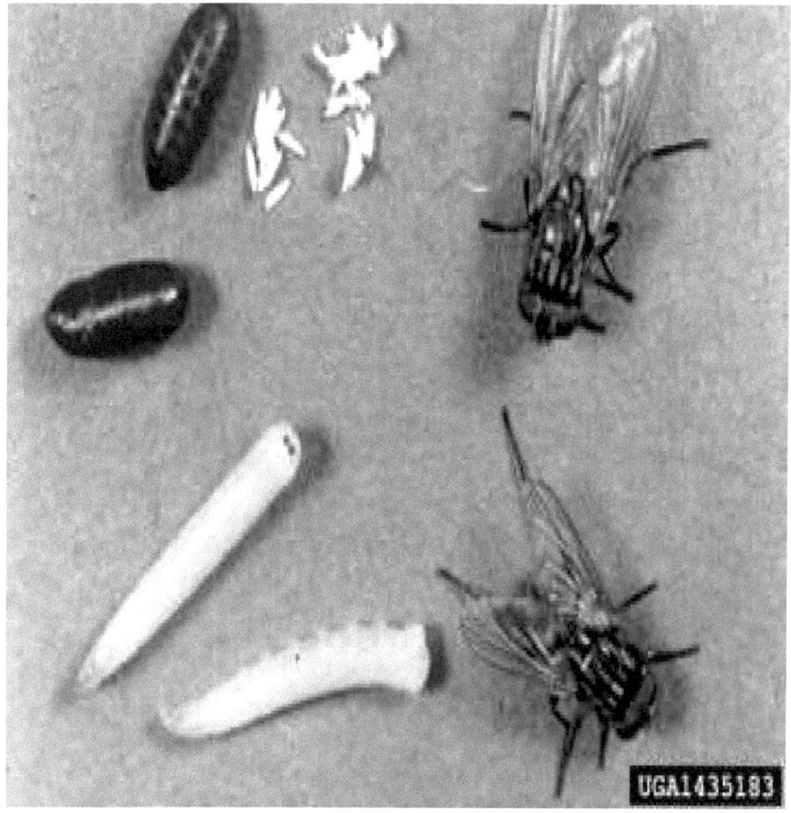

Fig. (17): The life cycle of flies

The egg stage

Whiteflies are laid in clusters containing 150-200 eggs in dark, secluded areas of the body, particularly around natural openings like the eyes, mouth, ears, nose, and anus, as well as in wounds and folds of the body behind the ears and beneath moist clothing.

The eggs, which are oval and white or yellow in color, typically hatch into larvae within a few hours or 1-2 days, depending on the temperature.

The larval stage

The body larvae have a tapered front, back, and mouth, with ventricles located at the back of the body. The larvae are white to yellow and go through three stages before reaching adulthood. The larval growth period lasts 12-15 days, depending on temperature and fly species.

The first stage larvae are 2 mm long, second stage larvae are 2-9 mm long, and third stage larvae are 9-22 mm long. Although larvae of different stages may vary in size and number of respiratory openings, they are generally similar in appearance. Larvae feed on body tissues.

Once fully developed, they cease activity, move away from the body, and bury themselves in the soil. In cases where the body is in a confined space without access to soil, larvae may form pupae on the body.

The Pupal stage

The Pupae can be found buried 3-5 cm beneath the surface of the soil. They are seen as the inactive phase of the flies prior to their full development, which occurs approximately 6-8 hours later.

The adult stage (fly):

After the fly exits the body of Virgo, its wings initially lie flat before stiffening and eventually allowing the fly to take flight .

Subsequently, the male fly will mate with the female fly, who will then deposits her fertilized eggs .

Rearing Fly Larvae

The larvae of flies are gathered from the scene of the crime and kept alive in the laboratory for breeding until their full development is necessary in order to classify the type of flies accurately at a later stage.

Flies are provided with various food sources such as beef, pork, or containers lined with tin foil or wax paper filled with industrial soil to absorb excess moisture.

Alternatively, fly larvae can be fed industrial foods like dry cat food, which does not produce odors and makes cleaning easier.

As the larvae mature, the color of the eggs changes from white with yellow hues to light red, dark red, light brown, and finally dark brown to black.

Once the larvae reach the pupal stage, their skin darkens and hardens. At this point, they should be separated from the industrial soil and transferred to individual containers for the emergence of the flies from the breeding cages.

When raising larvae for forensic purposes, it is important to avoid using industrial environments as they contain different components than natural food sources like meat, which could produce inaccurate results.

Pupae collected from crime scenes should be stored in specialized containers in the laboratory until the flies emerge, allowing for accurate species classification.

Effect of Environment on Forensic Insects

- **Climate seasons**

Insect behavior changes with the varying climatic seasons each year, influenced by fluctuations in temperature and humidity levels. Typically, insects that are commonly found on the body are more active during the warmer summer months.

- **Geographical location**

The activity of insects is influenced by varying geographic distributions, which in turn impact environmental conditions.

- **The soil**

The species of insects on a body and their behavior can be influenced by the body's location on the surface or bottom of the soil, hindering insects from reaching a body buried below the soil except for fly larvae which can reach bodies within a foot depth.

Bodies positioned above the soil surface are more appealing to various insect species.

- **The light**

Flies are insects that are active during the day and rest at night, making them diurnal creatures.

If a body is left out at night, flies will not be attracted to it until the next day.

• **Water**

The body being in the water stops fly insects from being attracted to it, whether on the ground or in the air, like beetles and flies, but it does attract aquatic insects.

• **Closed spaces**

Having the body indoors prevents flies from coming and laying eggs on it.

•**Clothing**

Having clothes on the body can lead to the creation of environmental conditions that attract and facilitate the multiplication of flies. The garments hold onto bodily fluids, such as sweat and other secretions, which serve as a source of moisture for flies to lay their eggs and for the larvae to develop on the body.

• **Drugs and toxins**

Medications accelerate the development of maggots on the body, but the poisons ingested by the victim hinder the growth of the larvae.

• **Body size**

Some species of insects are more attracted to small animals like mice, while other insects are more attracted to larger animals.

• **Cover the body with paint or grease**

Applying any type of paint on the skin acts as a barrier that prevents insects from being attracted to the body.

• **The dead body**

The carcass hanging in mid-air is more exposed to air, leading to a faster drying process, which makes it more appealing to insect species that are different from those attracted to a body buried in the ground or left on the surface.

It is important to mention that the soil beneath a suspended or buried body attracts various types of insects due to the fluids released from the body that collect in the soil, altering its pH and making it suitable for different insects. However, once the soil dries up, the insects originally present in the soil become active.

Entomological Evidences

Initially, it is important to gather plant samples from the crime scene before proceeding to collect insect samples or relocating the body, and before transporting it to the morgue, as mentioned earlier.

Studies have shown that burying or concealing the body in soil can decelerate the decomposition process compared to when the body is left on the soil surface. This is attributed to the involvement of bacteria and insects in the decomposition process on the surface, combined with the lower temperatures beneath the soil surface.

The density of the soil covering the body impacts the behavior of insects and their preference for specific species on the body. Research has indicated that a thicker layer of soil above the body can hinder the presence of flies flying above the surface, as certain flies need direct access to the body to lay eggs.

It has been noted that smaller flies may burrow into the soil above the body in order to lay eggs directly on it. Beetles can also burrow into the soil to reach a body buried at a depth of 1 m to feed on the remains. Her performance in the competition was outstanding.

Forensically Important Beetles

The beetles have Coleoptera wings, which have mouth parts that are segmented and sensory antennae made up of 11 segments. Their bodies are covered with a hard shell, known as elytra, which covers their membranous hind wings. Their legs are adapted for burrowing and some for swimming.

The larvae of beetles have a strong head capsule and a long, cylindrical body. Their mouthparts consist of mandibles and the prolegs on their abdomen are rarely present in larvae.

The pupal stage of beetles occurs in a chamber made of soil or within a cocoon in some species, from which the fully grown beetle emerges later on.

Some beetles prefer to inhabit the bodies of large animals, while others choose smaller hosts such as rats.

There are also social beetles where the mother beetle cares for her young larvae until they reach adulthood, protecting them from other insects that may prey on them.

1-The types of Forensically Important Beetles

- **Skin hide and larder beetle**

These beetles belong to the dermestidae family and their life cycle lasts between 20-50 days, which is influenced by the temperature of their surroundings.

They are classified as stored grain insects and are identified by the abundance of long bristles on both sides and the back of their bodies, which they use to vibrate when they sense danger or when they are touched or threatened with death (Fig. 18).

Fig. (18): Skin hide and larder beetle

• Clowen beetle

The histeridae follows this beetle. This beetle is easily recognizable by its unique name. Its distinctive shape is interesting. It is very small in size and has a shiny black color.

Both larvae and adult insects feed on fly larvae and other insects. The beetle first appears during the stage of body swelling, and remains present until the beginning of dry season, as shown in Fig. (19).

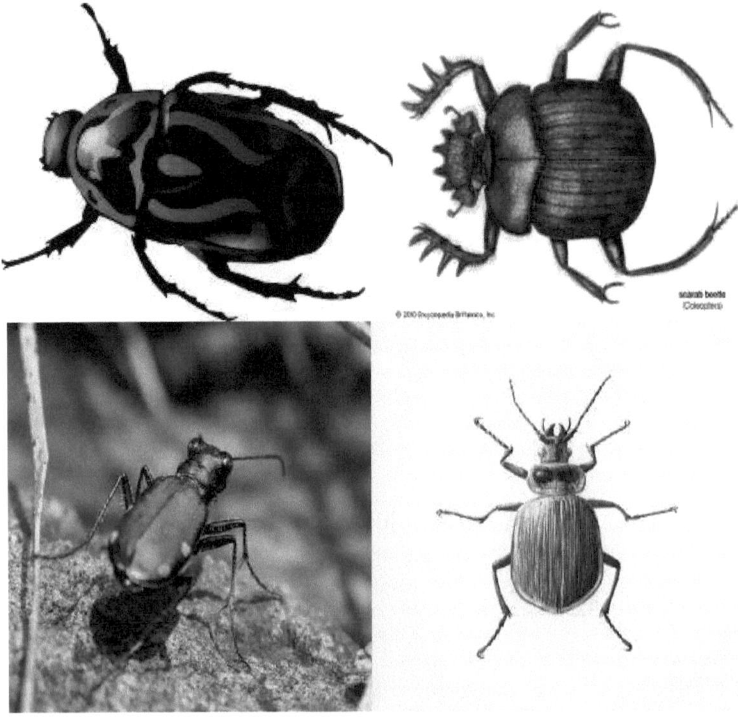

Fig. (19):Clowen beetle

- **Carrion beetle**

This particular beetle is often found alongside the Silphidae family, typically on bodies in various stages of decay. However, there have been instances where they have been observed on fresh bodies or bodies in the process of decomposition in certain areas, even without the presence of others.

It has been noted that certain beetles are influcnced by the size of the corpse. Some species tend to gravitate towards larger bodies, while others are seen with beetles on the bodies of smaller animals like rats.

The beetles have a unique shape, characterized by their sharp and curved bodies, with sensor horns that have thicker rings towards the tips.

These beetles are large and robust, with orange markings on the edges of their wings. Some are black, with abdomens consisting of six rings.

Fig. (20): Carrion beetle

•Checkered or Bone beetle

This particular type of beetle, which belongs to the Cleridae family, can be found on the body during different stages such as bloating and even drought. The presence of these beetles on the body can vary from one area to another, and they have a bright color that resembles medium and thick hair.

It is important to note that these beetles can result in body deformities that look similar to bullet wounds, as they create holes in the body when they feed on it. This can lead to confusion when trying to determine the cause of death. Additionally, the larvae of this species feed on the larvae of flies.

This species is often seen following Nerobia rufipes, commonly known as the red-legged ham beetle, Fig. (21).

Fig. (21): Checkered or Bone beetle

- **Dung beetle**

This type of insect belongs to the Scarabaeidae family and is known for living in tunnels beneath the ground, making it difficult to detect during crime scene investigations.

The occurrence of these beetles is often linked to the availability of suitable food rather than the stage of decomposition of the body at the crime scene. This is shown in Fig. (22).

Fig. (22): Dung beetle

- **Trogid beetle**

The Trogid beetles, which are medium-sized and pale brown with a rough back and wings that may have bristles, have legs crossed for drilling and long sharp claws. Their antennae end in flat rings. They are commonly found on small carcasses in the advanced stages of decomposition where they feed on fur, feathers, and dry remains.

These beetles are known for their ability to play dead (thanatosis) when startled or captured, as shown in Fig. (23).

Fig. (23): Trogid beetle

•Rove beetle

The Staphylinidae are known for their high level of mobility and energy. Their life cycle typically lasts 7-10 days, during which they eat fly larvae. They are drawn to a body shortly after death and stay with it as it decomposes, Fig. (24).

Fig. (24): Rove beetle

- ## Ground beetle

This beetle belongs to the carabidae family and is a nocturnal insect that is active at night and seeks shelter during the day, making it seldom seen. It goes through a life cycle of around one year, with the adult beetle living for 2-3 years, Fig. (25).

Fig. (25): Ground beetle

- **Nitidulidae beetle**

This particular species follows over 2,500 species of beetles. The beetles are known for their dark color and average length of 4-12 mm. They have oval bodies with elytra wings that are shorter than their body length. These beetles closely resemble the moving beetle, but can be distinguished by their antennae that resemble a racquet shape.

They are often attracted to fruit and rotting vegetables, while some are also drawn to decaying carcasses. These beetles thrive in moist environmental settings, as shown in Fig. (26).

Fig. (26): Nitidulidae beetle

Life cycle and Rearing of Beetles

Beetles undergo a complete metamorphosis during their life cycle known as Holometabolous.

The life cycle of a beetle begins with an egg, which is typically round or oval, followed by 3 to 5 larval stages, varying depending on the species of beetle. The larvae then develop into pupae before emerging as fully formed insects.

Beetles can either bury themselves in the soil or larvae can create a protective casing using soil components to enter the pupal stage (Figure 27).

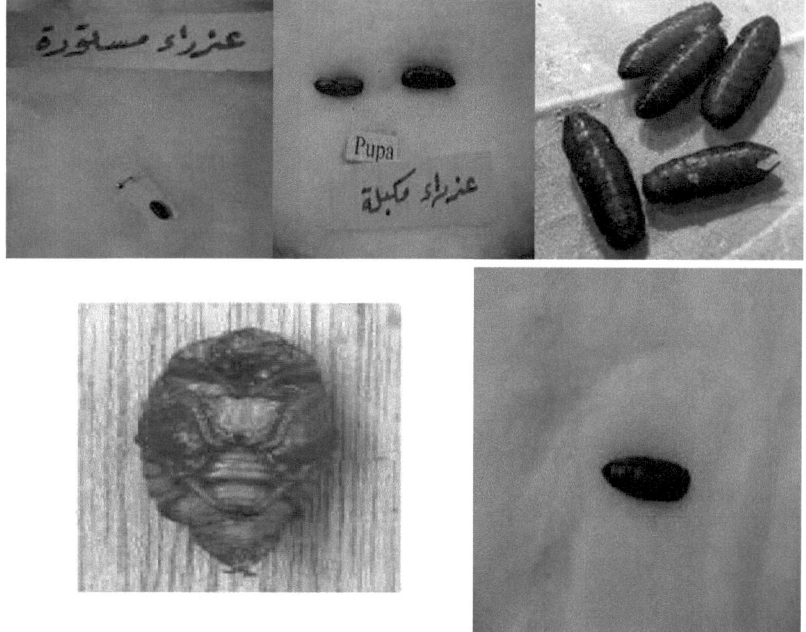

Fig. (27): Different from insect pupae

The beetles' life cycle duration varies based on their species.

The period from egg to full development in the Staphylinidae mobile beetle is approximately 7-10 days.

Carabidae beetles have a life cycle of about one year, with the entire insect living 2-3 years on average.

The number of larval stages in some beetles is not fixed and is influenced by environmental factors.

For instance, dermestidae beetles undergo nine larval stages.

These beetles typically have one generation per year.

The pupae of Dermests sp. take 2 to 9 weeks to develop.

During unfavorable winter conditions, some beetles can go into dormancy as pupae.

Identifying the various stages of beetle larvae is challenging, necessitating alternative methods.

Beetles are typically reared in plastic containers with a layer of soil and crumpled leaves for hiding.

Unilateral rearing may be required to prevent overcrowding.

To feed the beetles, frozen vara carcass or pieces of beef or pork can be provided.

Breeding vessels for the beetles are kept in a dark location like a nursery to promote the completion of the life cycle.

The Role of Aquatic Insects in Forensic Investigations

Despite the fact that only 3% of insect species have life stages in water, 95% of small invertebrate animals reside in water. The most crucial aquatic insects with significance in criminal investigations include:

•Stoneflies

The mature nymphs of this insect inhabit cold, swiftly flowing water. This insect is known for quickly hatching eggs, but the nymphs grow slowly and reach maturity after approximately a year, going through 12-22 stages of molting. Nymphs can be found living in the water at depths of several meters and have a lifespan of up to five weeks.

While the full-grown insect is typically active during the day, some species are active at night and rest during the day. The presence of these fly nymphs on the body in the water indicates that the body has been in the water for some time, allowing other predatory insects to survive by preying on the fly nymphs in nature's cycle, Fig. (28).

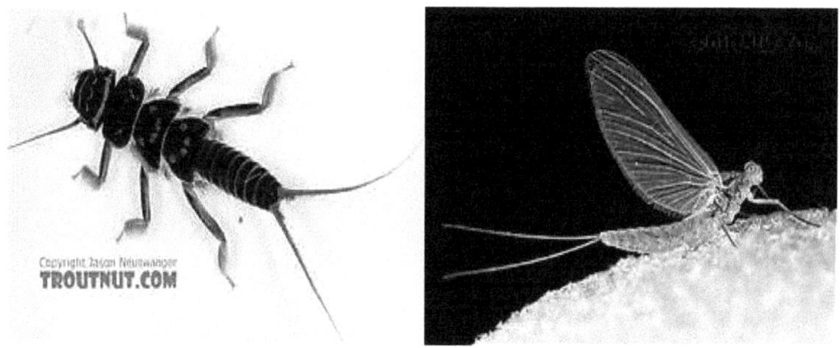

Fig. (28): Stoneflies

• May flies

The complete phase of the fly occurs for a brief period lasting a few days, during which the female lays 100-400 eggs. The incomplete stage is distinguished by a four-stage process in water.

The life cycle of the fly is diverse, with the first stage being the swimming phase, followed by a crawling and climbing stage. The third stage is described as oblate, while the fourth stage involves drilling into the bottom and completing 12-22 molts over 3-6 months before reaching maturity.

Nymphs feed on submerged organisms and can live up to three years, while adult insects are aerobic and typically seen around waterways on June nights. Fig. (29) depicts this phenomenon.

Fig. (29): May flies

• True flies

Over half of the identified aquatic insects exhibit this behaviour of following the flies, residing in various types of water bodies, including those that are polluted, and at different levels. The larvae of these flies consume algae and fungi, with traces of these flies' incomplete spores discovered on submerged creatures.

The larvae phase of these flies lasts for a period ranging from two to multiple years, contingent on the surroundings, while the adult insect is diurnal and nocturnal in its habits.

Clothes Moths

Tracking the types of butterflies clothing two groups: Tinedae and Oecphoridae and the most common types:

The Common moths, *Tineol bissellilla*

This particular butterfly is one of the most common types of insects, with a brown or golden color and measuring 5-8 mm in length. The larvae are white and typically feed on natural textiles like cotton and wool, especially those contaminated with sweat and urine. Additionally, they consume substances containing creatinine, such as bird feathers and hair, but do not feed on industrial textiles like nylon and polyester.

It is important to note that these larvae primarily feed on mammal body hair, with some even found inside the bosom. They begin to inhabit the body hair after hatching from the eggs and start weaving silk threads. The butterfly clothing lays eggs on leftover corpses, once the flies' larvae have completed their role in decomposing the body.

The larvae of this butterfly feed on both animal and plant material, as well as any food particles on the hair. They are typically the last insects to appear on a body during the final stages of decomposition. Once the larvae are fully developed, they create a thin wall made of fiber and foliage to enter the pupal phase. This phase prevents the butterfly from emerging, allowing it to complete its life cycle. It is important to mention that the butterfly does not feed, preferring darkness and flying away to hide when disturbed, Fig. (30).

Fig. (30): Butterfly clothes are common

The Case – bearing clothes moth, *Tinea pellionella*

This kind of mattress is like the Tineidae, which is less harmful than the previous type and is considered a textile variation. The caterpillars of this moth create a silk tube with added materials around the larva, as shown in Figure 31.

Fig. (31): The Case – bearing clothes moth, *Tinea pellionella*

The tapestry moth, *Tricophaga tapetzella*

Tineidae is the family of butterflies that are known to be attracted to and infest coarse clothing. This particular species of butterflies is highlighted in Fig. (32).

Fig. (32):The tapestry moth, *Tricophaga tapetzella*

The brown moth, *Hophmannophila pseudosprella*

The Oecophoridae species prefer this particular type of bedding. They devour animal and plant remnants found in bird nests and also consume textiles and stored items.

Acarology in Criminological Investigations

The group of mites and ticks known as Balakarus inhabit various environments and there are approximately 45,000 known species. It is believed that there may be up to half a million species that have not yet been described or classified, indicating that acarology lags behind entomology by around 50 years (Zaher 2011).

Different species of mites have varying lifestyles, with some feeding on plants causing economic harm, while others infest stored foods like dry cereals and cheese. Some prey on insects such as beetles and bees, while ticks and mites feed on animals, birds, and humans.

Certain species of mites, like dust mites and spider mites, have forensic significance in criminal investigations, known as Forensic Acarology. The Egyptian researcher "Rasmy" established this science in 2007 following the development of forensic entomology.

Despite their small size, some mites like Demodex folliculorum can have a significant impact on human health, particularly affecting the eyebrows and eyelashes by living in the hair follicles on various parts of the body. These mites can also cause eye inflammation (blepharitis) by transferring bacterial microbes.

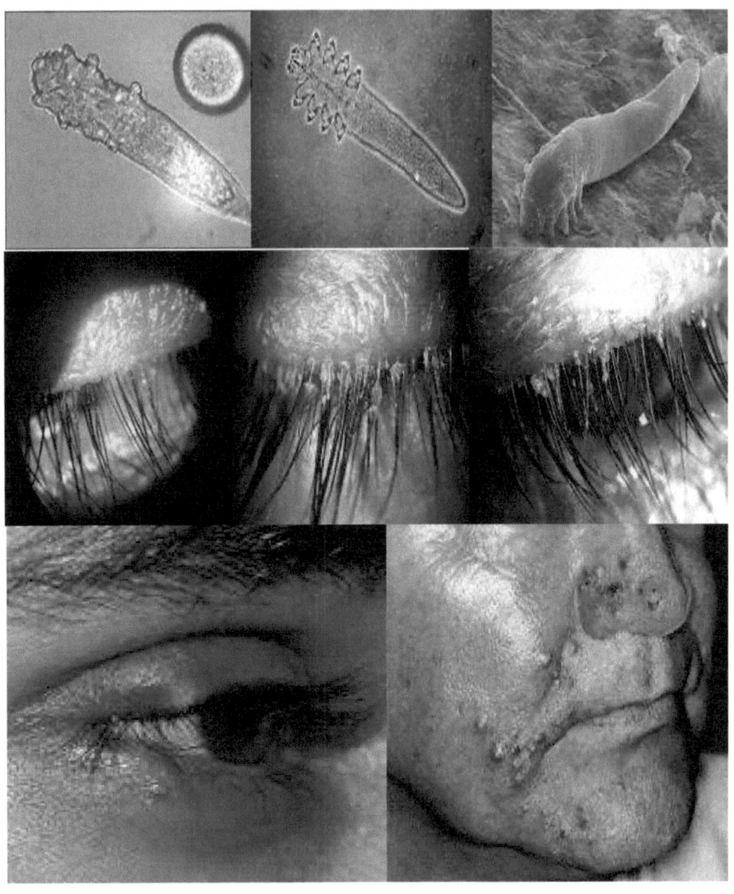

Fig. (33):*Demodex follculorum*

It also leads to various skin conditions and diseases in the human body, including skin infections and the transmission of diseases such as Rickettsia, known as Typhus, Friction Scrub Typhus, and other types that can affect the human body and live on the skin, causing diseases like scabies which is caused by the scab mite, Sarcoptes scabiei. This results in the spread of scabies among large groups of people, especially in poor areas and slums where health conditions are not suitable.

The spread of scabies among children, the elderly, and displaced individuals indicates a lack of care and negligence towards their well-being.

It has been observed that many people have some type of parasite living on their bodies, as evidenced by the presence of these parasites on the bases of eyelashes and in microscopic examination.

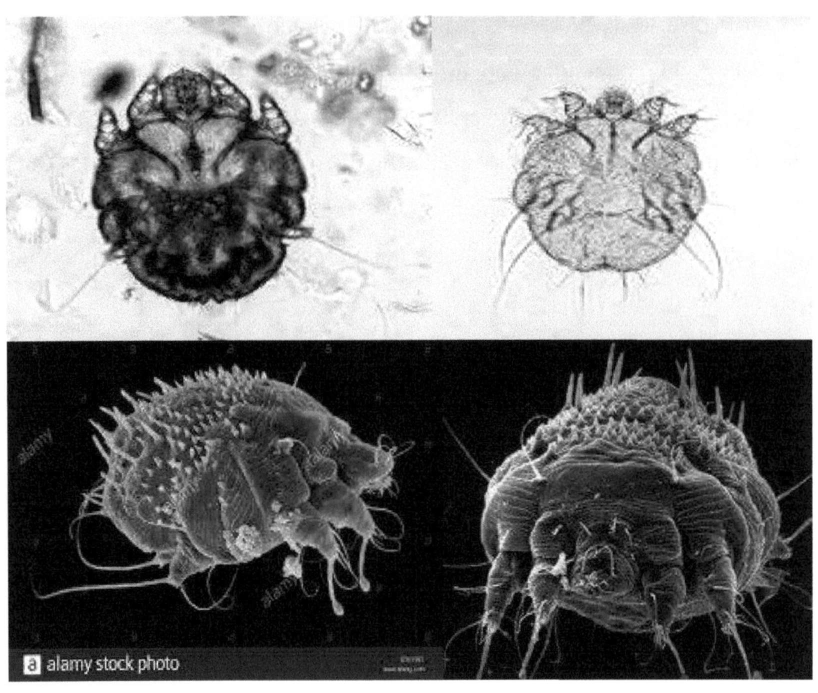

Fig. (34): The scab mite, *Sarcoptes scabiei*

The mites from the Demodex genus inhabit the sebaceous glands located around the nose and cheeks. Evidence of these mites can be found in fatty substances extracted from this area and observed under a microscope. Additionally, there are other mites, known as house dust mites, that live in dust, mattresses, and pillows, feeding on shed hair and causing respiratory issues like bronchial asthma, particularly in children.

One of the most common allergic diseases affecting the chest worldwide is caused by dust mites. To treat or prevent this condition, it is essential to eliminate the sources of dust mites, expose bedding to sunlight to kill the mites at different stages, and regularly clean living spaces where dust accumulates, creating a suitable environment for mites to thrive.

It is important to note that after a person dies, other types of mites and spiders of forensic significance may inhabit the body. These organisms often first arrive by hitching a ride on flies or through air currents, feeding on fly larvae and eggs, and reproducing quickly.

As the human body starts to decay, additional species of mites and spiders begin to infest the remains, feeding on fungi and other mites until the final stages of decomposition. French scientist Pierre Megnin elucidated this process in his renowned 1890 book "de Cadavers," highlighting how arthropods such as insects, mites, and spiders infiltrate the body in waves of eight sequences.

"The initial phase involves the mite species Macrocheles, which consumes fly larvae. During this stage, the mite is the primary or sole element of the sixth vector, indicating the level of body decomposition, followed by Tyroglyphidae and Oribatidae mites. Even after the body is removed from the crime scene, they may persist on the body or in the soil beneath it.

In 1894, Mignen was the first to utilize mites to estimate the time of death by studying the mite Tyrophagous, which lives in stored materials. He accurately determined the death time of a child whose body was concealed in a French home, dating back two years. Consequently, the current house residents were exonerated, while previous residents faced charges.

It is worth mentioning that mites and spiders have expanded their role beyond determining time of death, assisting in solving crimes like drug trafficking. In 2005, American scientist Starkby revealed that marijuana seized in the United States with spider traces only found in Colombia implies the drugs originated in Colombia before being smuggled.

Apart from aiding criminal investigations, some spiders can cause human fatalities due to venomous bites, leading to cell destruction. Species like the brown recluse, black widow, red widow, brown widow, and funnel-web spider are large and cause serious nerve injuries. Certain spiders and mites are pathogenic to humans, while others assist in criminal investigations to uncover crimes, Fig. (35, 36& 37).

Fig. (35): The brown recluse

Fig. (36): The black widow

Fig. (37): some spiders

Smell of Death

Descendants of the blowfly, known as the "flies of death," from the Calliphoridae family can be attracted to dead bodies from distances as far as several kilometers. This attraction is due to the post-mortem period immediately following autopsy, during which the body may be affected by various factors such as leprosy.

Once the flies have reached the site of the body, they typically seek out suitable locations to lay their eggs, often burrowing into the soil to reach the body. Research has shown that bodies buried at shallow depths of one foot or less in the soil can still attract flies, as their larvae are capable of penetrating the soil and reaching the body to feed on the tissues.

Coffinflies from the Ohridae tribe have been observed digging into the soil above buried bodies to lay their eggs directly on them, while Flesh flies from the Sacrophagidae family may lay their larvae directly on the body or on the soil surface above it to transport the larvae.

The presence of flies on the soil surface can sometimes indicate the presence of a body in that location within the past year or so. Beetles, such as the Burying beetle from the Silphidae family and the Staphylinidae rover beetle, have been observed to detect body odor from the soil surface and dig down to depths ranging from 10 cm to 100 cm., Fig. (38, 39 & 40).

Fig. (38): blowflies

Fig. (39): Flesh flies

Fig. (40): Burying beetle

DNA analysis of insects

1- DNA Analysis as Forensic Evidences

In 2001, Walse demonstrated that the DNA or physical trace of a person could be found in fly larvae that had consumed the body's tissues. The results of the analysis confirmed that the DNA found in the fly larvae matched that of the body tissue they had fed on.

In 2006, Szalanski and his team managed to extract human DNA from bed bugs that had fed on human blood. This breakthrough enabled forensic entomologists to utilize this technique in solving crimes and determining the perpetrator.

And for example:

A victim's body was DNA analyzed, revealing a match with larvae found in the intestines during a murder in Italy in 1989. The larvae were collected from land stored in a rural house.

The perpetrators successfully moved the body from the warehouse to another location to hide it before the police arrived to search the house. However, the DNA analysis confirmed that the DNA source was from a single corpse.

Investigators then disclosed that the victim's body had been concealed in the house, leading to the identification and prosecution of the criminal.

When collecting larvae for DNA analysis from the crime scene, it is important to kill them immediately by placing them in boiling water or storing them in 75% ethanol alcohol.

If larvae are not killed promptly after collection, important evidence from the crime scene located inside the larvae's intestines might be lost when the larvae defecate.

DNA found in the intestines of mosquitoes could also help in crime detection and identifying the perpetrator or victim.

Lice or pubic lice can also assist in crime detection as they feed on human blood and can be found on the body even after death if the body temperature is suitable.

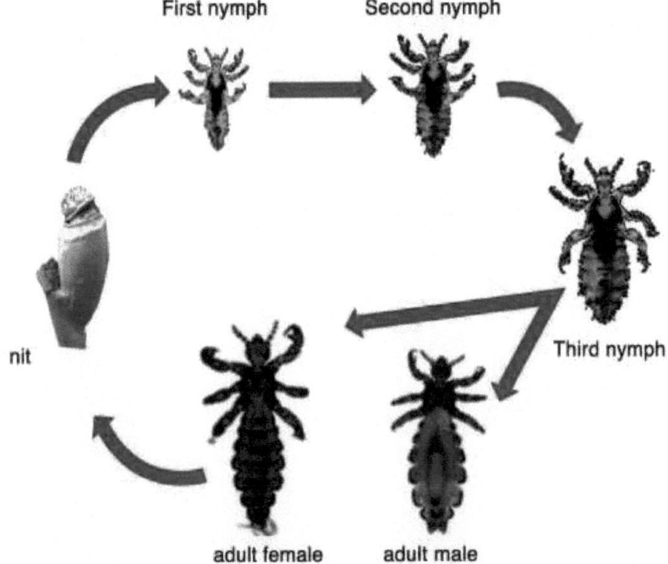

Fig. (41): Human lice, whose presence on the body is evidence that the date of death recently
After Dorthy Genard, 2007

Simpson claimed in 1985 that fleas could revive within an hour of being removed from a submerged body if the drowning lasted no longer than 12 hours, as shown in Figure 42 .

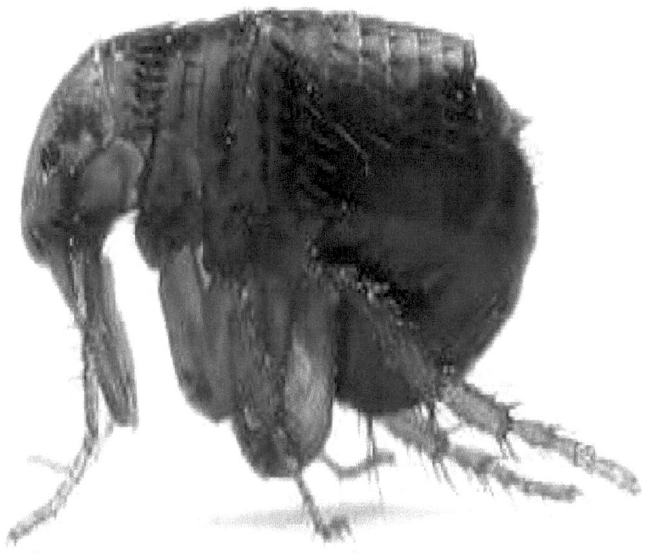

Fig. (42): The flea is an external parasite on the human body and can survive on the body of the shipwreck for about 24 hours (After Dorthy Genard, 2007)

2- Molecular Identification of Insects

Identifying the DNA of insects found on a body is a faster and more accurate way to determine the postmortem interval (PMI) than traditional methods that rely on the insect's physical characteristics and life cycle stages.

Example:

In 2007, the charred remains of a Taiwanese girl were discovered in a sugar cane field. The DNA analysis of fly larvae found at the crime scene matched the results of the morphological examination of flies produced from larval culture. The flies identified were of the Chrysomya megacephala species. To identify insect samples for classification based on their DNA, follow these guidelines:

1. Insect samples gathered from the crime scene can be preserved in 95% ethanol or in the refrigerator at zero degrees.
2. To minimize the chance of larval contamination with external DNA, larvae should be washed in a 20% bleach solution and acetone before DNA extraction, as these substances will not harm DNA components.
3. After washing the larvae, DNA and RAPD analysis can be conducted to determine the sequence of nucleotides on the chromosomes.

Before preserving the larvae for DNA analysis, it is important to first starve them and keep them in solutions or the refrigerator until their intestines are empty of waste and any external DNA that is not related to the insect species.

This is done to prevent any confusion between the insect DNA and other external DNA resulting from feeding on animal tissues.

Entomological Alteration of Bloodstain Evidence (Blood Spatters)

Violent acts often lead to bleeding, whether it is the blood of the victim or the blood of the person committing the crime. Therefore, the presence of blood at the crime scene can indicate what happened during the crime. Insects, particularly flies, are attracted to this blood and may leave markings on the walls after feeding on it, which can be further broken down by enzymes before the flies return to feed again and leave their mark.

It is important to note that blood stains left by flies on walls and near windows have a distinct appearance that sets them apart from blood stains caused by violence. These fly-generated stains are characterized by their light color, small size, and random tail-like patterns, differentiating them from human blood stains resulting from violent encounters, which are larger, darker, and tend to flow downwards, Fig. (43).

Hence, it is important for criminal investigators to differentiate between blood spots left by flies and bloodstains caused by violence at a crime scene in order to uncover the truth, identify the culprit, and avoid misunderstandings linked to the crime.

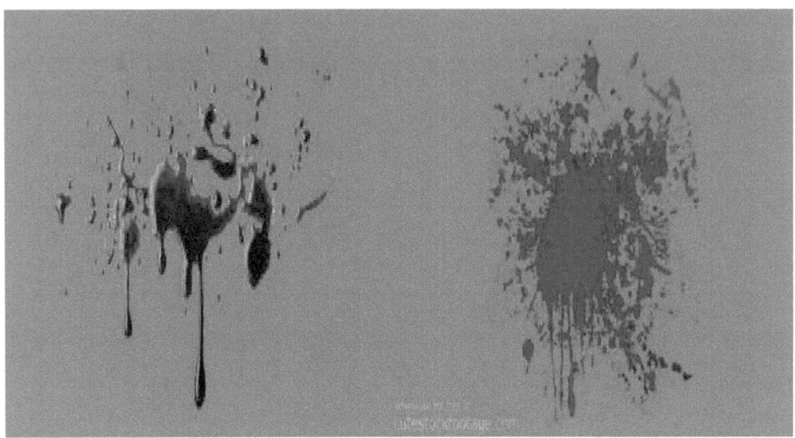

Fig. (43): On the right, human blood spots on a wall are characterized by a dark color and the tail of the spots goes down
While in the north blood stains due to the activity of flies crime scene and characterized by light color and the direction of the tail of the spots is random.

It should be noted that the cockroaches mixed with blood at the crime scene create a distinct trail that is different in shape than the trail left by flies. Cockroaches leave a path on each side, with a medium-length dark trail running between the light legs and the dragging of the abdomen.

Furthermore, the presence of additional blood spots in the vicinity of the victim's blood could indicate another injured individual responsible for the blood. It could also suggest the presence of another body at the conclusion of the blood trail.

The lack of evidence regarding the involvement of dream mites in this context could account for their small size and consequent small bloodstains that may be left at the crime scene, if they exist at all, rendering them invisible to the naked eye.

Example

In 1998, in Tallahassee, America, police instructed residents of a house to report a bad smell coming from a neighboring residence. Bloodstains were found at the entrance to the house and throughout the walls, with the body of the resident discovered in the bathtub surrounded by a pool of blood. Various blood patches were also found on the wall, door, and window.

The criminal insect expert concluded that the bloodstains on the wall were caused by flies. Empty medicine bottles and razor blades were found near the body. Upon further examination of the bloodstains resembling fly activity, it was determined that the death was not a murder but a suicide.

Myiasis

In 1974, James described the worm as an injury to human or animal tissue caused by fly larvae. This knowledge built upon Paton's findings in 1922, which indicated that living tissue could become infected at any stage of a fly's life cycle, with larvae being a significant causal factor.

The worm damages tissue by entering the body through natural openings such as the eyes, nose, ears, mouth, reproductive organs, and anus. It can also infect wounds where the necessary moisture is present for egg hatching and larval growth. The symptoms of the disease vary depending on where the fly larvae infection occurs. This can be seen in Figures 44 and 45.

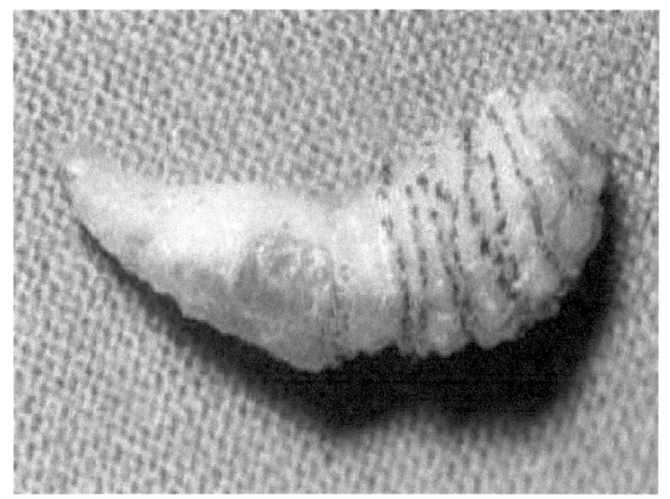

Fig. (44): ***Dermatobia hominis*** **(human botfly)**
Larvae causing the disease

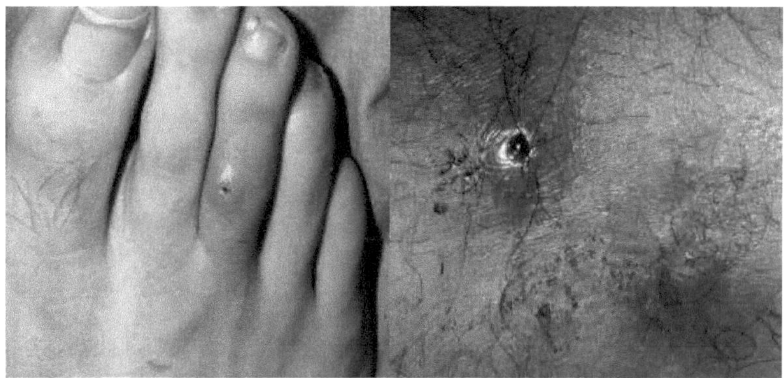

Fig. (45): Injury of flies to various wounds

Hand-treated by removing larvae and then cleansing wounds with a disinfectant with a broad-range antibiotic.

Botanical Evidences

Plant specimens found at the crime scene include various sizes such as Pollens, Fungi, Algae, flowers, flower petals, plant roots extending below the body, branches covering the body or fallen leaves.

Examining newly grown plants attached to the victim's hair or clothing may help determine the time of death or the crime scene location, while some plant specimens could implicate a suspect.

Before gathering insect samples from the body, it is essential to collect plant samples, document and photograph any changes in the crime scene caused by the perpetrator.

Plant leaves samples are stored in paper bags labeled with the collection date and other relevant information:

1- The frequency of this particular plant species in the vicinity of the location where the crime took place should be noted.

2- The hue of the plant specimen when it was gathered should be documented.

3- The closeness or attachment of the specimen to the deceased's body should be considered.

4- The classification of the specimen, whether it is a weed, a branch from a tree, or a flower, needs to be identified.

Upon collection, the plant sample must be analyzed by a botanist. Assessing these samples will provide insights into their ecological significance.

An expert in plant life can determine the environmental conditions based on these samples. It is advisable to include site-specific and plant-specific photographs of the specimens. Also, the botanical evidence related to the car of the culprit may be evidence of his conviction if proved to be the presence of the crime scene, as this may be considered evidence of the presence of the car in advance scene crime.

Changes to the body after death

1- Stages of Corpse Decomposition

After a person dies, enzymes in the body become activated, starting the process of decomposition called autolysis. As the body decomposes, it goes through five distinct stages:

- **Fresh phase Fresh (1-2 days)**

At the moment of death and before the body begins to decompose, it enters a stage where various flies are drawn to it. These flies lay eggs in the body's natural openings like the nose, ears, eyes, mouth, and genitals, as well as in any wounds where there is enough moisture for the eggs to hatch and larvae to develop.

- **Bloated stage (2-7 days)**

Ammonia, hydrogen, sulfur, and nitrogen cause the body to turn greenish-black, take on a balloon-like shape, and prompt the emergence of beetles while increasing the activity of larvae.

As a result, fluid starts to be released from the natural openings of the body, altering the environment around it to facilitate the growth and movement of different insects. During this stage, larvae reach their highest level of activity.

- **Decay stage (5-13 days)**

During this phase, the body bursts open and releases gases, leading to a reduction in size. The fly larvae feed on the remaining body parts until they eventually move to the soil on the tenth day after consuming all tissue. Beetles are also present and active during this period.

- **Post-phase stage (10-23 days)**

The extent of decay of the body is confined to the leftover body parts such as cartilage, bones, skin, and hair, leading to an increase in beetle activity along with a rise in the presence of predatory insects and parasitic fly larvae, as well as an increase in ant activity. At the conclusion of this phase, the population of beetles starts to decline.

2- The stage of the skeleton Dry remnants skeletal stage (18 days - to infinity)

Only the skeleton and hair remain in the body at this point, and the environment around the body starts to revert to its original state, including the presence of insects such as flies and spiders.

The last phase of the body's decomposition does not have a specific end point.

It should be noted that different organs decompose at different rates, which can help determine the time of death. The typical order of decomposition is as follows:

1- Organs such as the intestines, stomach, liver, and heart muscles are affected.
2- The lungs and airways are also impacted.
3- The brain undergoes changes as well.
4- The colon and bladder are part of the process.
5- Muscles throughout the body are involved in this.

6- The body goes through stages of stiffness within 36 hours after death known as Rigor Mortis.

7- After death, the body starts to decompose, leading to changes in body temperature depending on the surrounding atmosphere.

8- If the body is still warm and muscles haven't stiffened, then death likely occurred recently.

9- When the body is warm but muscles start to stiffen, it indicates that death happened 3-8 hours ago.

10- If the body is cold and stiff, it suggests that death occurred within 8-36 hours.

After the death period exceeds 36 hours, the muscles of the body relax and become rigid. Rigor Mortis disappears with the body cooler, depending on the Starkeby 2004 schedule, which helps determine the time of death if detected within 36 hours of occurrence, Table(3).

Table (3): Determine the time of death if detected within 36 hours of occurrence

date of death	Hardening of the body	Body temperature
The body temperature is less than 3 hours	Not rigid	warm
3 - 8 hours	rigid	warm
8 - 36 hours	rigid	cold
More than 36 hours	Not rigid	cold

3- Changes in the human body within one month of death Clark, et al., 1997

Table (4): Determine the time of death within one month of death

Time after death	The changes
Directly	Stop blood circulation - Breathlessness - Muscle relaxant - Paleness of color - Early blue under the surface of the body to deposit blood.
2 hours	The beginning of the hardening of the body - the beginning of the body cooler - blue body color - changes in the blood vessels of the eye.
4-5 hours	Blood clot - blue pattern
24 hours	Dryness of the cornea of the eye - Return of blood flow
48 hours	Back muscle relaxation
72 hours	Hair loss and nails
96 hours	Bacterial activity - changes in skin and skin
Days - Month	Greenness of body color - Body bulge - Exits gases - Exits fluids from the body - Start of the appearance of the Great Temple - Completion of the appearance of the Great Temple

The employment of insects in forensic examinations

Forensic entomology involves investigating the use of insects and other arthropods in criminal cases. [39] These insects and arthropods are typically found in decaying animal corpses or carrion. [40] By studying these insect colonizers, scientists can make estimations about the time of death, determine the movement of the corpse, identify the manner and cause of death, and potentially link suspects to the crime scene.[41]

In the 13th century in China, the first documented case of insects being used in a criminal investigation is mentioned in Sung Tzu's book about "the washing away of wrongs." After a farmer was found murdered in a field with a sharp weapon, all suspects were instructed to place their sickles on the ground. Only one sickle attracted blow flies to a small amount of blood that was not visible to the naked eye, leading to the confession by the murderer. [39]

The first use of forensic entomology in a modern court setting occurred in 18th-century France, where entomological evidence was used to acquit the current residents of a home where the skeletonized remains of a child were discovered. During the 18th century, Yovanovich and Megnin's study on insect succession on decomposing bodies laid the foundation for the field of forensic entomology. [40]

Arthropodes and their association with postmortem changes of the human body

Upon death, cells begin to die and enzymes begin to break down the cells through a process known as autolysis. The body then enters a state of decomposition. Bacteria in the gastrointestinal tract start breaking down soft tissues, emitting liquids and gases such as hydrogen sulphide, carbon dioxide, methane, ammonia, sulfur dioxide, and hydrogen.

The release of volatile molecules, known as apeneumones, from the decomposing body attracts insects. Scientists have been able to identify the specific volatile chemicals released during various stages of body decomposition. These molecules can alter the behavior of insects that are attracted to the decaying matter. [42]

According to research conducted by Crag et al. in 1950, it was discovered that sulfur-based compounds were responsible for initially attracting flies to the decomposing carcass. However, the presence of ammonium-rich compounds on the carrion is what triggers the flies to lay eggs on the decomposing remains. 43]

Smith (1986) identified four different categories of insects that can be observed on decomposing carrion. These include:

1. Necrophagous species that feed directly on the carrion
2. Predators and parasites that feed on the necrophagous species, including some schizophagous species that initially feed on the body and later become predatory
3. Omnivorous species that feed on both the carrion and other arthropods like ants, wasps, and certain beetles
4. Other species such as springtails and spiders that use the corpse as part of their habitat.

The first two categories are particularly significant in the field of forensic entomology, with the majority belonging to the Diptera (flies) and Coleoptera (beetles) orders. The succession of arthropods colonizing the carrion is influenced by the stage of decomposition. [40]

Insect species commonly involved in forensic investigations are usually true flies or Diptera. The most common types within this order include Calliphoridae (blow flies), Sacrophagidae (flesh flies), and Muscidae (house flies). Calliphoridae and Sacrophagidae can appear shortly after death, while Muscidae typically wait until the body begins to bloat. Calliphoridae adults are often shiny and have metallic colors such as blue, green, or black on their thoraxes and abdomens.

Sarcophagidae are medium-sized flies with black and gray stripes on their thoraxes and checkering on their abdomens. Adult Muscidae are typically 8-12 mm long with gray thoraxes and dark lines on their backs. They also have hair-like projections covering their bodies. These insects typically lay their eggs in natural body openings or wounds, which hatch into maggots.

Maggots are small, peg-shaped organisms with mouth hooks on one end for feeding. They go through three stages of growth, known as instars, before reaching full size. Once fully grown, maggots stop feeding and move to drier areas to begin pupariation, forming a protective encasement as they transform into flies. [44]

According to research conducted by K. Tullies and M. L Goff on carcasses exposed in a tropical rainforest, it was determined that the decomposition process could be categorized into five stages based on the physical appearance of the carcasses, internal temperatures, and the types of insects present. The first stage is known as the Fresh stage, which occurs during the first 1-2 days after death and ends when bloating of the carcass becomes visible. Autolysis takes place during this stage, but there are no significant morphological changes. Entomological data is more accurate than a medical examiner's estimation for determining the time of death after 24 hours.

Insects are attracted to the carcass within the first 10 minutes of death during the Fresh stage, but egg laying (oviposition) does not occur at this time. Cellular breakdown occurs without any visible morphological changes or odors, although the chemicals released during this breakdown attract insects, even in the early stages of decomposition. .[42]

Swollen stage (Days 2-7): Decay starts during this phase. The buildup of gases from anaerobic bacteria's metabolic processes causes the abdomen and the carcass to swell, giving it a balloon-like appearance towards the end. The combination of arthropod actions and decay raises the internal temperatures of the carcass. The highest numbers of adult Diptera are attracted to the carcasses at this stage. Diptera larvae in the first and early second instar stages are present by the fourth day. Additionally, several predators of Diptera larvae are found on the carcasses by the beginning of Day 2.

Decomposition stage (Days 5-13): The abdominal wall is pierced, leading to the deflation of the carcass, marking the end of the swollen stage. Internal temperatures rise 14 degrees higher than the surroundings and then drop, signaling the end of the decomposition stage. Strong odors are produced during higher temperatures and decrease with cooler temperatures. The carcass weight steadily decreases by the 10th day, with a conversion of carcass biomass to Dipteran larval biomass. The larvae then leave the carcass to pupate.

Advanced decomposition stage (Days 10-23): This phase starts once most of the Diptera larvae have left the carcass, leaving bones, cartilage, hair, small bits of tissue, and a significant amount of wet, sticky material known as byproducts of decay (BOD). The BOD is where most arthropod activity occurs during this phase.

Remaining stage (Days 18-90+): Bones with minimal cartilage and dried-up BOD characterize this stage. The transition from the advanced decomposition to the remaining stage is gradual, with decreasing populations of adult and larval Diptera. [45]

Steps in estimating the postmortem index with insect larvae

Insect larvae found on a deceased body can help determine the postmortem interval (PMI) for up to a month. Accurate identification of the species is crucial. Different species have varying growth rates and maturation processes. To estimate the PMI, the age of the larvae must be established. The age of the larvae can be approximated by measuring the length or dry weight of the oldest larvae and comparing it to standard data. The larvae's developmental pace is influenced by the ambient temperatures in their surroundings. Each stage of development requires specific temperatures, leading to each species having its own set number of accumulated degree days or hours for full growth. By knowing the thermal history of the larvae, a comparison can be made with temperatures at the location of death to estimate the PMI. The age of the first-generation adult flies can also indicate the duration since death. [45] These flies can be identified by their withered wings and small abdomen with a dull gray shade. When the insects that colonize a corpse in a particular area are known, a model of insect-colonization succession can be utilized to estimate the PMI.[46,47]

Using insect data for determining the site of crime

Differences in insect species found on decomposing corpses can vary depending on the habitat and environment. By closely examining the insects present, researchers have discovered that the species associated with a corpse in one type of habitat may differ from those found when the corpse is moved after death.

DNA analysis for species identification

The first crucial step in estimating the age of larvae is correctly identifying the species. Typically, species identification is carried out through morphological comparison, a method that necessitates specialized knowledge and can be quite time-consuming. To address this challenge, species identification can be accomplished by amplifying specific regions of the larvae genomes using polymerase chain reaction and comparing the results with reference data. [48]

Entomotoxicology

The maggots of the flies that consume dead flesh can store substances taken in by the deceased individual. Bodies that are highly decomposed or reduced to skeletons pose challenges in testing for toxic substances. In such cases, the maggots feeding on the body can be crushed and examined using methods such as thin-layer chromatography, gas chromatography, and/or mass spectrometry.

Various toxins can impact the growth stages of the maggots. Substances like cocaine and heroin present in the corpse can speed up the larvae's growth.

Toxic chemicals like malthione in the dead flesh can hinder insect colonization. .[48]

Insect Orders

Diptera (Flies)

Calliphoravomitoria

Luciliaampullacea

Chrysomyamegacephala

Chrysomyarufifacies

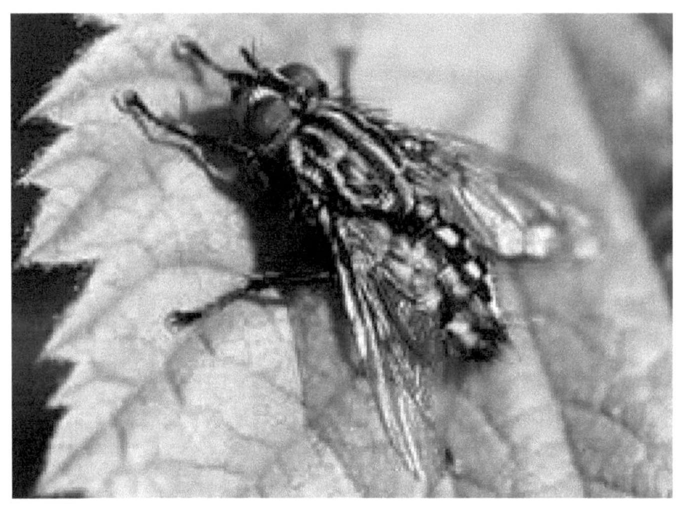

Sarcophagaalbiceps

Coleoptera(Beetles)

Muscasorbens

Calosilphasp.

Dermestesmaculatus

Gonocephalum sp.

Philonthuslongicornis

Saprinussp.

Conclusion

Criminal insect science has developed very rapidly over the past few years, and courts have begun to accept insects as a criminal guide.

On the level of the Arab world there is interest on the personal level of one of the legal doctors in Egypt, but by participating in a seminar in Syria under the supervision of Naif Arab University for Security Sciences shows that this science is not applied at the official level, in any of the Arab countries so far.

It is worth noting that Kuwait is represented by the Ministry of Interior.

The General Directorate of Criminal Evidence has established a technical committee to oversee the implementation of criminal insect science in the State of Kuwait and has developed a database of criminal insects for different Kuwaiti environments.

We hope that this science will be implemented and applied at the level of Arab countries to help uncover the various crimes and help to achieve justice.

Forensic entomology, a growing field in forensic sciences, involves the study of insects that feed on dead bodies. It has become a crucial tool in criminal investigations. Nowadays, forensic odontologists do more than just examining hard tissues. The rising involvement of forensic odontologists in criminal investigations as part of the forensic team highlights the importance of raising awareness about emerging sciences like forensic entomology and its role in forensics.

The capability to extract human and mammalian DNA from insects has significant implications in forensics, especially in terms of contamination and as a possible source of targeted DNA. While insects have been occasionally used in forensic investigations to transmit DNA, increasing awareness of this emerging field of forensic entomology would be advantageous for investigators. Research is necessary to determine the impact of various variables on insect-facilitated DNA transfer and its effects on forensic investigations.

The observed variation in amplification and genotyping success emphasizes the importance of identifying which components in insects and/or their artifacts are impacting the extraction and amplification efficiency, and how these impacts can be reduced. It would be beneficial to conduct more comparative studies to determine the best practices for sampling, preservation, extraction, amplification, and genotyping techniques, with the goal of creating a recommended protocol that enhances DNA recovery while preserving essential morphological characteristics to avoid interference with other entomological analyses, such as species identification .

Further exploration of the morphology of insect artifacts, especially those left by untested insects, could provide valuable insights, although testing every species of a specific insect is unnecessary. It has been acknowledged that attempting to document all potential artifact morphologies is impractical, and creating a definitive test for identifying artifacts would be more advantageous. In this regard, the ongoing development of an immunoassay (Rivers et al., 2018; D. B. Rivers et al., 2019) is a promising step in that direction .

The amount of forensically useful information attainable from insects, as well as the risk of DNA contamination from insect-mediated transfer, will rise with the increasing sensitivity of DNA detection methods and the improved DNA recovery from insects and their artifacts. Therefore, it is crucial to raise awareness among crime investigators and scientists about the significance of insects as carriers of biological material and DNA in forensic investigations.

Insectattractiontoandinteractionwithhumanbodyremainshasbeenwitnessedand evenusedforcenturies,andtodaymedicocriminalentomologyisbeingpracticedacrosstheworld.Thereisnodearthontheavailabilityofscientificliteratureonthistopicbutthenumberofqualifiedpracticingforensicentomologistscapableoutilizinginsectevidenceis exceedinglysmall.Indevelopedcountries,thisbranchofsciencehasreachedatanexcitingstageinitsevolutionastestimonybasedontheinsectevidenceandisroutinelyusedinthe courts;buttherearenotakersofthisbranchofappliedscienceinIndiaandotherdevelopingnations.Thereisastrongneedtoacceptandrecognizethedisciplineofforensicentomologyandtoencourageitsrelianceinthecourtoflawasbiologicalevidce,whichwillpresent immenseopportunitiestothequalifiedforensicentomologistsaswell.

REFERENCES

1. Joseph I, Mathew DG, Sathyan P, Vargheese G. The use of insects in forensic investigations: An overview on the scope of forensic entomology. J Forensic Dent Sci. 2011;3(2):89-91. doi:10.4103/0975-1475.92154
2. Forensic Entomology: The Use of Insects in the Investigation of Homicide and Untimely Death by Wayne D. Lord, Ph.D. and William C. Rodriguez, Ill, Ph.D, The Prosecutor, Winter 1989.
3. Madhu B, Anika S. Postmortem Interval Estimation of Mummified Body Using Accumulated Degree Hours (ADH) Method: A Case Study from Punjab (India). J Forensic Sci & Criminal Inves. 2016; 1(1): 555552
4. Aly H. Rasmy, The humans lie but the spiders do not lie: An overview on forensic acarology, Egyptian Journal of Forensic Sciences,Volume 1, Issues 3–4,2011,Pages 109-110, ISSN 2090-536X, https://doi.org/10.1016/j.ejfs.2011.07.001.
5. Alejandra Perotti, M., Lee Goff, M., Baker, A.S. et al. Forensic acarology: an introduction. Exp Appl Acarol 49, 3–13 (2009). https://doi.org/10.1007/s10493-009-9285-8.
6. Amendt J, Krettek R, Zehner R. Forensic entomology. Naturwissenschaften. 2004 Feb;91(2):51-65. doi: 10.1007/s00114-003-0493-5. Epub 2004 Jan 16. PMID: 14991142.
7. Brian D. Morice, et al. "Necrophagy In Honey Bees (*Apis Mellifera L.*); A Forensic Application of Scent Foraging Behavior." Journal of the Kansas Entomological Society, v. 92,.2 pp. 423-431. doi: 10.2317/0022-8567-92.2.423
8. Honey bees and their honey could be a big help in solving police cases", George Mason University, Jan 2022.

9. **Anderson, G. S. 1995.** The use of insects in death investigations analysis of cases in British Columbia over afive year- period Canadian Society for Forensic Science Journal, 28:277-292.

10. **Bass, J. 2006.** *Carved in Bone: A Body Farm Novel.* New York, NY: HarperCollins, 2006.

11. **Benecka, M. 2004.** Arthropods and corpses. In: Tsokos M. (ed), Forensic Pathology Reviews, vol. 2. Humana: Totowa, NJ: 207-240.

12. **Benecka, M. and Lessig, R. 2001.** Child neglect and Forensic entomology. Forensic Science International, 120: 155- 159.

13. **Benecka, M. and Barksdale, L. 2003.** Distinction of bloodstain patterns fly artefacts. Forensic Science International, 137: 152- 159.

14. **Benecke M. 2001.**brief history of forensic entomology, Forensic Science International, 2001; 214.

15. **Byrd. J. H. and Castner, J. L. (eds) 2001.** Forensic entomology: the utility of arthropods in legal investigation. CRCpress: Boca Raton, F. L.

16. **Carloye, L. 2003.** "Of Maggots and Murder: Forensic Entomology in the Classroom," *The American Biology Teacher,* Vol. 65, 2003.

17. **Catts, E. P., and N. H. Haskell.2000.** *Entomology and Death, a Procedural Guide*. Clemson, SC: Joyce's Print Shop, 2000.

18. **Cheng, K. O. 1980.** Cases in the history of Chinese trials [ENGLISH TRANSLATION OF Zhe yu gui jian bu] lu shin China. In entomology and law, Greenberg B.and Kunich J. C. (eds) Cambridge University Press: Cambridge, 2002.

19. **Colyer, C. N. 1954.** The coffin fly, *Conicera tribialis* Schmitz (Dipt., phoridae). Journal of the society for British Entomology, 4(9): 203-2006.

20. **Dillon, L. and Anderson, G. S. 1996.** Forensic entomology: the use of insects in death investigations to determine elapsed time since death in interior and northern British Columbia regions. Technical report TR-03-96. Canadian police Research Centre: Ottawa, Ontario.
21. **Gannon, R.1997.** "The Body Farm," *Popular Science*, September 1997, p. 77.
22. **Genard, D. E. 2007.** Forensic entomology: An Introduction. Jon Wiley and Sons. Ltd, 244Pp.
23. **Gennard D.2007.** ; Forensic Entomology An Introduction, Wiley 2007.
24. **Goff, M. L. 1993.**Estimation of postmortem interval using arthropod development and successional patterns. Forensic science Review, 5(2): 81-94.
25. **Goff, M. L. and Flynn, M. F. 1991.**Determination of postmortem interval by arthropod succession: a case study from the Hawaiian Islands. Journal of Forensics. 36: 607-614.
26. **Goff, M. L. 2000.** *A Fly for the Prosecution*. Cambridge, MA: Harvard University Press, 2000.
27. **Greenberg, B. and Kunich, J. C. 2002.** Entomology and the law: flies as forensic Indicators. Cambridge University press: Cambridge.
28. **Hall. R. D. 1990.** Medico criminal Entomology. I cats, E. P. and N. H. Haskell. Eds., Entomology and Death: A proceeding guide. Clemson. Se: Joyce's Print Shop. 1-8.
29. **Haskell N & others; 2008.** Entomology & Death: A procedural guide, Clemson, SC; East Park Printing, 2008, 2nd edition.

30. **Megnin, J. P. 1894.** La Faune de Cadavers. Application de l,entomologie a la medicine legale. Encyclopedie scientifique des Aide. Memoires. Paris: Masson et Gauthier Villaars.

31. **Rasmy, A.H. 2007.** Forensic Acarology: Anew area of insect and death: An overview on Forensic Entomology, First Arab Intr. Forensic science & Medicine Acarology Conf. Proc. 1-6, Riyadh, S. A. An overview on Forensic Acarology. Egyptian J. of Forensic Sciences, 1(3-4): 109 – 111.

32. **Rasmy, A.H. and Gesraha, M. A. 2015.** Criminal research and insects (The role of insects in detecting crime) Library of Egypt Deposit number2015 / 11017 Pp 130.

33. **Sachs, J. S. 1998.** *Corpse*. New York: Basic Books, 2001. Sachs, J. S. "A Maggot for the Prosecution," *Discover*, November 1998, p. 103.

34. **Schoenly, K. G., N. H. Haskell, D. K. Mills, C. Bieme-Ndi, K. Larsen, and Y. Lee.2006.** "Using Pig Carcasses as Model Corpses," *The American Biology Teacher,* Vol. 68, 2006.

35. **Staerkeby, M. 2005 .**Ultimate guide to Forensic Entomology: Introduction wep only essay. URL< HTTP://FOLK.uIO.nO/MOSTARKE/FORENS-ent/introduction.shtml>. Accessed 5Jan. 2004.

36. **Tantawi, T. L. , El-Kady, E. M. Greenberg, B. and El-Ghaffar, H. A. 1996.** Arthropod succession on exposed rabbit carrion in Alexandria, Egypt Journal of Medical Entomology, 33(4): 566-580.

37. **Turcketto, M., Lafisca, S. and Constantini, G. 2001.** Postmortem interval (PMI) Determined by study of *Sarcophagous biocenoces*. Three cases from the province of venice (Italy). Forensic science International. 102:28-31.

38. **Zaher, M. A. and Hassan, M. F. (eds). 2011.** Harmful phytophagous mites. Faculty of Agric. Cairo Univ. press, 216Pp.

39. Catts EP, Goff ML. Forensic entomology in criminal investigations. *Annu Rev Entomol.* 1992;37:253–72.

40. Amendt J, Krettek R, Zehner R. Forensic entomology. *Naturwissenschaften.* 2004;91:51–65. [PubMed] [Google Scholar]

41. Sukontason K, Narongchai P, Kanchai C, Vichairat K, Sribanditmongkol P, Bhoopat T, et al. Forensic entomology cases in Thailand: a review of cases from 2000 to 2006. *Parasitol Res.* 2007;101:1417–23.

42. LeBlanc HN, Logan JG. Exploiting Insect Olfaction in Forensic Entomology. In: Amendt J, Goff ML, Campobasso CP, Grassberger M, editors. *Current Concepts in Forensic Entomology.* Netherlands: Springer; 2010. pp. 205–21.

43. Ashworth JR, Wall R. Responses of the sheep blowflies Lucilia sericata and Lxuprina to odour and the development ofsemiochemical baits. *Med Vet Entomol.* 1994;8:303–9.

44. Goff ML, Lord WD. Entamotoxicology;a new area of forensic investigation. *Amer J foren med pathol.* 1994;15:51–7.

45. Tullis K, Goff M L. Arthropod succession in exposed carrion in a tropical rainforest on O'ahu Island, Hawai. *J. Med. Entomol.* 1987;24:332–9.

46. Schoenly K, Reid W. Dynamics of heterotrophic succession in carrion arthropod assemblages: discrete series or a continuum of change? *Oecologia (Berlin)* 1987;73:192–202.

47. Davies L. Species composition and larval habitats of blowfly (Calliphoridae) populations in upland areas in England and Wales. *Med Vet Entomol.* 1990;4:61–8.

48. Chen WY, Hung TH, Shiao SF. Molecular identification of forensically important blow fly species (Diptera: Calliphoridae) in Taiwan. *J Med Entomol.* 2004;41:47–57.

Websites

1. www.amonline.net.au/insects/insects/resources. htm#forensic; many good links
2. www.forensicentomology.com/introduction.htm; a very complete exposure to forensic entomology
3. http://research.missouri.edu/entomology; more links www.nhm.ac.uk/nature-online/life/insects-spiders/webcast-forensicentomology/forensic-entomology.
4. html; video explaining forensic entomology

About Authors

Assist. Prof. Dr. Senaa Abdullah Ali Al- jarjary

Department of biology, College of Science, University of Mosul. Iraq.

Email: Sensbio23@uomosul.edu.iq

Mobile: (+964)7701871500

Prof. Dr. Raed Salem Ahmad ALsaffar

Department of Forensic Evidence, College of science, University of Mosul, Iraq.

Email: raesbio2@uomosul.edu.iq

Mobile: (+964) 7701634647

Prof. Dr. Mohamed Abdel-Raheem Ali Abdel-Raheem

Professor of Entomology (Biological Control)

Pests & Plant Protection Department

Agricultural and Biological Research Institute

National Research Centre.

33rd ElBohouth St., Dokki, Cairo, Egypt.

Email: abdelraheem_nrc@hotmail.com,

abdelraheem_nrc@yahoo.com

Mobile: (+2) 01155527583 - (+2) 01009580797

Prof. Dr. Mohamed Abdel-Raheem Ali Abdel-Raheem, Prof. of Entomology (Biological Control), Pests & Plant Protection Department, Agricultural and Biological Research Institute, National Research Centre, Cairo, Egypt. Published publications (267) papers (80), Books and Book chapters (187), (Scopus) h-index (10), Citations (258), (Google Scholar) h-index (16) , Citations (765), (Web of Science) Publications (9), h-index (4), Citations (37), Research Gate, h-index (13), Citations (539), Reviewers in International Journals (125), Reviewed articles in Biological Control (495), Editor in Chief in Journal (6), Editorial board in Journals (32), Associated Editor in Chief in Journal (5), conference Attended (27), workshop Attended (292), Symposium Attended (176), Forum Attended (4), Prize (4), Attended Others (30), Projects As PI and member (19), Training courses For Agric. Engineering (20), Training courses For Students of University (8), attending Trainees courses (19), TV & Radio (20), Supervisor on Ph.D. thesis (1), Committee Ph.D. thesis (2), and member in scientific Foundation (13).

National Research Centre. 33rd ElBohouth St., Dokki, Cairo, Egypt. Email Address: abdelraheem_nrc@hotmail.com, abdelraheem_nrc@yahoo.com, ma.abdel-raheem@nrc.sci.eg, Mobile Phone: (+2) 01155527583 – (+2) 01009580797

- https://www.scopus.com/authid/detail.uri?authorId=55220661700
- http://scholar.google.com/citations?hl=en&user=gBtssEgAAAAJ
- https://orcid.org/my-orcid?orcid=0000-0001-9240-064X

- https://www.webofscience.com/wos/author/searchAbdel-Raheem, Mohamed Abdel-Raheem Ali - Web of Science Core Collection
- https://www.researchgate.net/profile/Mohamed_Abdel-Raheem4/publications?sorting=recentlyAdded
- http://livedna.org/20.16018
- http://www.resarcherid.com/rid/R-6264-2017
- https://www.webofscience.com/wos/author/record/R-6264-2017
- https://www.youtube.com/attribution_link?a=NsH2xGv9mY8&u=%2Fwatch%3Fv%3Dsb46rBG8-Mc%26feature%3Dshare
- https://youtu.be/Ub8djW0tggo
- https://m.facebook.com/story.php?story_fbid=492907069288338&id=100076011247165
- https://youtu.be/dOujUAZmTdw
- https://m.facebook.com/story.php?story_fbid=492907069288338&id=100076011247165
- https://fb.watch/gPGrDuzZFX/
- https://www.youtube.com/watch?si=4UfyQaeHVLPiVR5J&v=nXV4NT-no7o&feature=youtu.be
- https://www.youtube.com/watch?v=sRH44W6TbcQ
- https://youtu.be/uq9Qj3AEBHk?si=ymaX06huG2AXJESu

Printed by Books on Demand GmbH, Norderstedt / Germany